Ham Radio's Technical Culture

Inside Technology

edited by Wiebe E. Bijker, W. Bernard Carlson, and Trevor Pinch

A list of books in the series appears at the back of the book.

Ham Radio's Technical Culture

Kristen Haring

The MIT Press
Cambridge, Massachusetts
London, England

© 2007 Massachusetts Institute of Technology

All rights reserved. No part of this book may be reproduced in any form by any electronic or mechanical means (including photocopying, recording, or information storage and retrieval) without permission in writing from the publisher.

Every effort has been made to contact those who hold rights for all materials reproduced here. Any rights holders not credited should contact the publisher so a correction can be made in the next printing.

MIT Press books may be purchased at special quantity discounts for business or sales promotional use. For information, please email special_sales@mitpress.mit.edu.

This book was set in Stone Serif and Stone Sans on 3B2 by Asco Typesetters, Hong Kong, and was printed on recycled paper and bound in the United States of America.

Library of Congress Cataloging-in-Publication Data

Haring, Kristen.
Ham radio's technical culture / Kristen Haring.
p. cm.—(Inside technology)
Includes bibliographical references and index.
ISBN-13: 978-0-262-08355-3 (alk. paper)
ISBN-10: 0-262-08355-8 (alk. paper)
1. Amateur radio stations—History. 2. Radio operators—United States—History.
3. Hobbies—Social aspects. I. Title.
TK9956.H3626 2007
384.540973'0904—dc22 2006046597

10 9 8 7 6 5 4 3 2 1

for my brother, Keith, with abiding love and respect

Contents

Prologue

Every night thousands of men retreat to radio stations elaborately outfitted in suburban basements or tucked into closets of city apartments to talk to local friends or to strangers on the other side of the world. They communicate by speaking into a microphone, tapping out Morse code on a telegraph key, or typing at the keyboard of a teletypewriter. In the Internet age, instantaneous, long-distance, person-to-person communication seems ordinary. But amateur radio operators have been completing such contacts since the 1910s.

The hobbyists often called "hams" initially turned to radio for technical challenges and thrills. As the original form of wireless technology became more reliable and commonplace in the 1930s, ham radio continued as a leisure activity. This book examines why men in mid twentieth century America operated two-way radios for recreation and how the hobby shaped social and technical encounters. It primarily concerns the period after radio broadcasting became routine and before personal computing did. The hobby is still widely practiced, with more than 680,000 hams in the United States in 2000—more than ever before. While there may be many points of continuity between past and present ham radio, what follows is a historical analysis based on evidence from the 1930s to 1970s and aiming only to interpret events of that era.

To become an amateur radio operator required considerable skill, machinery, and time. The first hurdle was obtaining a license from the Federal Communications Commission (FCC) by passing a written examination of electronics theory and radio regulations and a hands-on test translating words into and out of Morse code. Once he earned an FCC-assigned call sign, the hobbyist next had to either buy or build the equipment for his home station. A two-way radio station needed a transmitter, to generate and send out signals, and a receiver tunable over the particular frequency range the FCC reserved for amateurs. Successful communication depended on additional gear—from an antenna and headphones to diagnostic equipment and tools such as a voltmeter, oscilloscope, and soldering iron. This stockpile of devices demarcated the hobby space or "shack," which took its name from the "radio shacks" that housed communication equipment on board ships and for military field operations.[1] Though shacks often were relegated to the basement, attic, garage, or other unrefined parts of a home, hams prized these territories set apart from domestic activities, completely devoted to radio. Postcards confirming individual contacts usually decorated the walls, along with any awards and the hobbyist's FCC license. A large desk provided comfortable operating conditions, and shelves of manuals and magazines served as a technical reference library. For construction and repair projects, ideally a shack also contained a workbench. Assorted spare parts might be strewn about or stored neatly in bins, depending on the hobbyist (see figure P.1). In periods of tinkering with equipment that could stretch on for months, the ham resembled the stereotypical lone inventor. Then a flip of a switch and a spin of a dial brought the many voices of hobby radio rushing into the shack.

Dialing through the band of frequencies set aside for amateur radio unleashed a cacophony. Layers of voices, in different languages, competed with the staccato tones of Morse code, whose rhythm and strength varied according to the style of the human sender and the power of his transmitter. Only with precise tuning and some luck could a clear signal be isolated. Ham radio operators used streamlined language and repetition of key phrases to cut through static and background chatter. According to the standard procedures for initiating a dialogue, a call by licensee KB3DF requesting to talk with anyone available would be spoken as, "CQ CQ CQ, this is KB3DF calling CQ. Kilowatt bravo three delta foxtrot, calling CQ CQ CQ." The code for a general call ("CQ") might be modified to "CQ DX" to elicit a response from a distant station ("DX" being radio jargon for long-distance operating) or be followed by the call sign of another hobbyist when answering a specific person's CQ. During the rush of a contest or when conditions were poor, conversations stuck to a dry exchange of data

Figure P.1
Bill Higgins, W0YDB, in his ham radio shack in 1968. Confirmation postcards, awards, a map, and his FCC-issued license covered the walls. Photograph printed with his permission.

about station location and reception strength. Under other circumstances, two hams might speak at length about their lives and hobby involvement, even if meeting for the first time. A contact ended with sending "best regards," couched in the code phrase "73," and declaring "over and out" before recording the date, time, operating frequency and power, and the other party's license number in a log book that was subject to FCC inspection.

For all its technical trappings, ham radio thrived on social interaction. It differed from amateur broadcasting such as pirate radio and from pastimes focused on listening to commercially broadcast or shortwave radio because it included both transmission and reception. This produced real-time conversations (not necessarily comprehensible to non-hams), and random meetings "on the air" occasionally grew into friendships that continued by letters and further discussions via radio. Hobbyists who lived near each other gathered in clubs or met informally for "eyeball contacts." United by their recreational application of radio technology and distinguished by their electronics skills, hams sometimes described themselves as a "technical fraternity." The number of amateur license holders in the United States—around 100,000 in the early 1950s, twice that by 1960, and 375,000 in 1979—was sufficient to sustain an intricate social network and a profitable niche industry. Yet radio hobbyists remained a minority and celebrated this as a sign of technical superiority.

Hams spoke of themselves as democratic and open to all who made the effort to learn radio theory and operation. The mid twentieth century hobby radio community, however, was remarkably homogeneous. The following chapters document the subtle but intentional process by which the community became intensely masculine—an overwhelming majority of hams were male, and the hobby culture played up the manliness of radio activities. Other demographic characteristics emerged from multiple contributing factors. That ham radio operators generally belonged to the middle and upper socioeconomic classes partly reflects how expensive it was to participate in the hobby. Men also improved their financial standing by using skills gained in the hobby to launch lucrative electronics careers. The education level and occupations of hams can be seen either as following from their class status or as following from their technical passions and then altering their class status. On average, a radio hobbyist completed more years of schooling than the non-hobbyist—after World War II this usually included some college—and he was far more likely to hold a job in a technical field. The military recruited hams for their radio skills, and military service gave hams further technical training and eased access to higher education under the GI Bill. In this way, technical inclination, recreation, education, skill, and employment reinforced one another to the extent that it is impossible to separate cause from effect when questioning their relationship to class. The racial homogeneity of hams lacks explanation beyond its socioeconomic connections. Statements of racial and ethnic identification among hams were rare, but polls confirmed that the white faces filling radio magazines accurately represented the ham population. The community discouraged all internal divisions except geographic ones, denouncing religious and ethnic radio clubs as "political" and therefore a potential hindrance to smooth relations with federal regulators.[2]

While the prospects for international communication created a great deal of excitement and anxiety about mid century ham radio, Americans dominated the hobby. In 1960, when more than 200,000 amateurs in the United States held licenses, Great Britain had the second most hams, with only around 9,400. Roughly half the world's countries then had less than 25 registered hobbyists each, and only 16 countries had more than 1,000 hams.[3] These figures—compiled by a hobby magazine to inspire respect for the difficulty of contacting foreign operators—speak to the political, economic, and technical position of the United States as well as to the American enthusiasm for technology at mid century. Increased Cold War funding for military technology and the championing of electronics for strategic, productive, and recreational purposes supported the hobby. At the same time, the climate of secrecy and isolation prevalent during this period of global tension meant that hams who sought private, international ties provoked suspicion.

Nowhere did ham radio technology appear more out of place than in the suburbs of the 1950s. On a typical evening, families inside houses arranged in orderly developments, with neatly manicured lawns, gathered around television sets to watch light entertainment. A local ham disrupted this scene visually and electronically. Neighbors wrinkled up their noses at the strange-looking antenna mounted atop the hobbyist's house or attached to a tall tower poking out of his lawn (see figure P.2). Even when this "contraption" was out of sight, it was hard to forget about the ham down the block. Amateur radio operators, broadcast radio listeners, and television viewers all enjoyed recreation based on the wireless transmission of electrical signals. Interference occurred if a ham's transmission strayed from the frequencies designated for amateurs, or if a television or radio receiver picked up signals outside the broadcast frequency range. Without realizing it, a hobbyist chatting on the airwaves might produce a series of beeps and buzzes on the channel where his neighbor had hoped to find the night's baseball game on the radio. Sometimes pieces of a ham's conversation could be heard clearly on nearby television sets, drowning out the broadcast's sound and ruining the picture, too. These bizarre occurrences raised the ire and piqued the curiosity of those living close to radio hobbyists. It was a time when signs at some military bases warned, "Talk means trouble—Don't talk," when Americans feared outside influences

Figure P.2
Especially when mounted on a tall tower, a ham antenna conspicuously marked the home of a radio hobbyist. Printed with the permission of photographer Robert Walsh, WB3AMY.

and obsessed about the threat of communism.[4] So why was the neighborhood ham sitting down in his basement talking to Russians? One hobbyist's wife reported that "all his friends quit speaking to him because he's ruined their favorite television programs" and claimed that her whole family had "become suspect and is shunned by polite society."[5] Compounding the social rebukes, municipalities charged many hams with zoning violations related to "unsightly" antenna towers, and the FCC imposed operating restrictions and fines on amateurs caught interfering with commercial broadcasts.

The apprehensiveness of non-hobbyists about amateur radio was understandable. While most Americans witnessed the formidable technical realm of the military-industrial complex from the sidelines, hams were right in the thick of it. The hobby had an intimate relationship with electronics, the showpiece technology of the period from World War II into the 1970s. This helped hams extend their leisure pursuits into hi-tech civilian and military careers. And hobby radio organizations persuasively lobbied the FCC to maintain a portion of the airwaves for amateurs because two-way radio had value as a strategic technology. All of which begged the question of whether anyone should be tinkering with such powerful devices in his spare time. Outsiders alternated between teasing hams for choosing an odd hobby and revering hams for their technical expertise. Hams were geeks with an adventurous side, who could be counted on to solve (and cause, sometimes) electrical problems; they were, in this sense, precursors to computer hackers.

Hams deliberately set themselves apart by developing a community and culture tied to radio technology. They articulated technical values, goals, and practices different from those of non-hams and used adherence to this way of thinking to judge group members. That is, radio hobbyists formed their own "technical culture," a culture built around and establishing an ideology about technology. Studying a community defined by beliefs about technology highlights the creation and implications of technical culture. I hope that my presentation of the notion of technical culture through the example of ham radio will stimulate investigation into other technical communities and ultimately offer insight into the formation and function of the technical cultures that are so familiar to us that we take them for granted.

Ham radio existed within a larger category of technical hobbies. I point this out not to downplay that several qualities made it a truly unique pastime. Hams engaged in communication on a global scale, using equipment that rarely was seen outside of the military, subject to strict state regulation—the last of these aspects following directly from the first two. The consideration of radio hobbyists in the context of hobbyists who raced miniature airplanes, modified motorcycles, and built personal computers demonstrates the ways in which ham radio was exceptional as well as what it had in common with other activities. The book begins by defining the category of technical hobbies and explaining the motivations and experiences shared by people who took up technology for leisure. Later chapters trace how hams formed a community around a technology and crafted a particular image of ham radio, how the culture of hobby radio affected the market for equipment, and the consequences that practicing ham radio had in hobbyists' relationships with employers, with the state, and with their families.

This is a text-based history. I did not have access to old audio tapes of on-air conversations, though over the years I have spent many hours casually observing in a ham shack. Given the huge number of longtime radio hobbyists, I contemplated conducting interviews as part of my research. But the rich documents hams produced allowed me to avoid the challenges of oral history, such as selecting representative informants and interpreting their comments in light of the fact that decades had passed since the events described. There is a small secondary literature on amateur radio, focused almost exclusively on the 1910s and 1920s. Susan Douglas perceptively chronicles early ham radio in *Inventing American Broadcasting*; several of the numerous histories of radio briefly mention the first hams; and Clinton DeSoto's 1936 *Two Hundred Meters and Down* provides an insider's technical history of amateur radio.[6] These books gave me a picture of a quite different hobby than existed at mid century and allowed me to isolate potential roots of that difference, which helped guide my research through the primary literature. Most radio clubs published informal monthly or quarterly newsletters packed with local and personal information. Handbooks sold to hobbyists and the manuals that manufacturers included with equipment reveal the style of technical lessons (often interlaced with social lessons) pitched at hams. To understand ham radio's connection to the state, industry, and the public, I consulted government documents, trade literature, and general magazines and newspapers.

Hobby periodicals deserve a special introduction because they formed such a vital source of evidence for this study. *QST* and *CQ* were the leading monthly hobby magazines with national circulations in the 1940s and 1950s. *QST* (the title is code for "calling all members") debuted in 1915. As the organ of the American Radio Relay League, the main amateur radio promotion and lobbying organization, *QST* claimed to set out the "official" positions on hobby matters, though it had only a self-declared authority. *QST* tended to distance the League from any controversy and to present a united front, even when none existed among hams. *CQ*, a less authoritarian commercial publication begun in 1945, did not shy away from printing multiple points of view. These magazines were joined in 1960 and 1968 by two more independents that became popular, *73* and *Ham Radio*. Together these periodicals reached a majority of hams. In the early 1960s, *CQ*, *QST*, and *73* had combined subscriptions that exceeded the number of licensed hobbyists in the United States by about 20%.[7] This subscription tally double counts individuals who received more than one of the magazines but also includes libraries and clubs, where many individuals would have read a single issue. Perusing the articles, advertisements, editorials, and letters in hobby publications, I found the topics that mattered most to hams and the spirit that enlivened their pursuits.

No general account of the hobby can adequately convey the personal stories of the roughly one million Americans who operated amateur radios over the course of the twentieth century. I expect that this book will prompt diverse hams to speak up about their own experiences and how those may break from my analysis. If I succeed at least in convincing readers of the relevance of technical recreation, the addenda offered by hams should gain the attentive ear of non-hobbyists, including future scholars.

Before proceeding, I feel obliged to address the standard question of why amateur radio operators are called "hams." The hobby community generally agrees that the origin of the nickname will remain a mystery, all the while debating the matter in good humor. Proposed derogatory explanations for the term that circulated in the hobby literature include that early wireless enthusiasts were known for "hamming it up" on the air and that professional telegraphers berated amateurs for having a "ham fisted" clumsiness with telegraph keys. Other common legends suggest that a shortening of "amateur radio" to "am. radio" shifted to "ham radio" for ease of pronunciation, that a club station before the days of FCC licensing took one initial from each of its three members' names as the call sign "HAM," or that hobbyists who operated out of abandoned smokehouses referred to these buildings as their "ham shacks." Whatever the etymology, hobbyists played to the name's obvious negative connotation in facetious recipes for cooking hams and jokes about amateurs' piggishness. The pride with which hobbyists accepted the peculiar moniker reflects their eagerness to identify with amateur radio.

1 Identifying with Technology, Tinkering with Technical Culture

I set out to write a book about amateur technical practices. I planned a chapter on ham radio, one on model rocket construction, others on computer hacking and modifying motorcycles. From the start, I sensed that there was something special about ham radio, so I began my research there. My hunch proved correct, but as I read through stacks of amateur radio club newsletters and technical handbooks, I was drawn back to thinking of ham radio as one pastime among many whenever hams alluded to amateur photography or antique car restoration. I kept asking what united these activities and what differentiated ham radio enough to justify focusing on it. The result is a book on ham radio that begins with a chapter on technical hobbies.

Until the 1880s, "hobbies" included all manner of personal obsessions. Americans in the late nineteenth century adopted a different meaning. Since then, strictly speaking, the term "hobby" refers only to pursuits distinguished by their association with values such as productivity, educational enrichment, thrift, and the structured use of time. Contrasted to idle recreation, hobbies were thought to keep participants busy with activities that led to personal betterment. A magazine for builders of mechanical models in 1925 listed "thought and care, infinite patience and perseverance" as important moral lessons that came "coupled with skillful workmanship."[1] The belief that select leisure activities fostered positive attitudes and character traits led social service agencies to promote hobbies during the Great Depression in an effort to maintain an industrious work ethic despite the high rate of unemployment. Pastimes appealing to a wide range of interests, like coin collecting, needlepoint, gardening, do-it-yourself household projects, and scrapbooking, all qualified as hobbies. Depending on personality as well as on class-related factors such as amount of disposable income and free time, individuals might pursue these activities casually or—as in the pre-1880s meaning of "hobby"—obsessively.[2]

Ham radio fits the strict definition of a hobby. Several aspects of the ideology that the ham community developed with reference to technology can be traced to values fundamental to all hobbies. Being a hobby also links ham radio to other pastimes in a way that raises useful questions. Within the diverse category of hobbies, participants and scholars alike identify subcategories such as craft hobbies or collecting hobbies. I have come to think of ham radio as belonging to a subcategory of technical hobbies, and clarifying which activities should be grouped under that heading helped me appreciate why people turned to technology for recreation and how this affected attitudes toward technology. This analysis draws on the slim secondary literature that addresses technical hobbies and on my own primary historical research into hobbies that seemed closely related to ham radio. I look forward to refining my initial attempt at classifying technical hobbies as scholars examine the many worthwhile topics in this area that remain open for investigation.

To count as a technical hobby in my description, the productive recreation essential to hobbies must require some technical understanding or skill beyond simply how to operate a technology. Also, each technical hobby has as its focus some machine or apparatus, but this characteristic is not sufficient for a hobby to be termed technical. The definition includes hobbyists with a wide range of expertise and involvement, though not necessarily of equal status within the hobby community. (Technical hobbyists evaluate technical ability just as all sorts of hobbyists pass judgment about the talent and commitment of fellow participants.) The intersection of technology with hobbies generated a rhetoric that embraced certain modes of twentieth century technical work as fun. It is to emphasize this heritage that I use the term "technical hobby" as opposed to "amateur technology."[3]

Naming some examples will sharpen the category. Building ultralight airplanes, working toward increasing the accuracy of sound reproduction in the playback of recorded music, recreational computer programming, miniature engine construction, and creating "chopper" motorcycles all can be classified as technical hobbies. Not every hobby that uses technology is a technical hobby. While working with saber saw, chisel, router, mallet, and drill, a recreational furniture builder displays considerable technical expertise but technology does not typically motivate the pastime. Amateur photographers who use cameras as tools differ from the furniture builder in their intention to capture images by means of a particular technical process, whether manual or automated. Exploring limitations of the central tools, such as shutter speed and the sensitivity of film media, is one of the educational components that contributes to marking amateur photography as a hobby, especially in contrast to non-hobby photography. Hobbyists are deeply engaged with technology even if they keep their hands outside of machines. Like amateur photographers, many computer hobbyists modify hardware only minimally or not at all, instead focusing their tinkering on software.

Given variability of practice, there can be hobby and non-hobby versions of the same activity. To spend a Saturday replacing the head gasket on a truck displays the productivity and thrift of hobbies along with technical knowledge yet should not be called a hobby if performing repairs is a rare activity done only out of necessity. If, on the other hand, an old truck is kept around partly as a project, for the challenge of keeping it running or the satisfaction of restoring it to mint condition, it very well may be the center of a technical hobby. The determination would hinge on how the individual pursues, feels about, and characterizes the activity, and on how it is viewed by the community of hobbyists.

The technical activities mentioned here all involve machines. This is chiefly a consequence of considering only productive leisure pursuits and is not meant to suggest a simplistic equation of technology with machines. If pressed to define "technology," I also would include some processes and some objects that are not machines, like pasteurization and bridges. Still, I am operating with a classic definition of technology—as the physical application of scientific knowledge—that may sound perfectly standard to most readers but terribly outdated to some of my colleagues in technology studies. The tendency in scholarship of the last few decades has been to take a broader view and to label nearly any kind of skill or know-how as technology. This inclusivity, as I understand it, is an attempt to avoid the glorification of a category that otherwise could be perceived as masculine, capitalist, and Western/white. I prefer another route to achieving similar ends: instead of collapsing all categories of skill into the technical, I find it more revealing to expose how technology took on its specific sociopolitical identity and to question the valuation of technical skill over other kinds of skill. Merely classifying ham radio as a technical hobby and knitting as a non-technical hobby does not impart additional qualities to either activity, nor does it rank one above the other.

Technical hobbies are in fact largely practiced by men, but this was not inevitable. The image of ham radio as manly only resulted from the ongoing, deliberate efforts of ham radio operators. This process of masculinizing a technology is documented in this book. Documenting a reciprocal process, "how boys have historically been socialized into technophiles," Ruth Oldenziel presented the example of the building contest run by the Fisher Body Company, a supplier of auto bodies to General Motors.[4] That she and I each use cases of activities promoted to boys and young men to illustrate the forging of a connection between technology and masculinity points out the way early lessons contribute to the naturalization of gender. It also is possible that the construction of technology as manly within the lower-stakes leisure arena made it difficult to question this characteristic within the workplace, where it had more serious ramifications. So far, much of the concrete evidence we have about the gendering of technology comes from leisure studies. Hobbyists who collected sounds with electronic recording devices, to name an additional example, used sporting language to associate their pastime with more traditional men's activities.[5]

Unfortunately, historical demographic data on technical hobbyists is difficult to gather. Club membership lists and the names and photographs in the hobby literature offer a guide to gender, but an unknown number of hobbyists never joined clubs, so the gender ratio within clubs or texts may not accurately reflect the gender ratio of hobbyists overall. The general consensus among casual and scholarly observers, however, is that men dominated technical hobbies. Two studies agree that audiophiles are men with above-average income and education levels.[6] Otherwise, we know precious little about the class, race, and other attributes necessary to compile a basic profile of technical hobbyists. Detailed examination of particular hobbies is needed to find this information.

Many individuals pursued multiple technical hobbies simultaneously, demonstrating that hobbyists were not fixated on single devices. That the Schenectady (New York) Photographic Society hosted a show of "the latest equipment and gadgets on the movie market" could have displayed a predictable affinity between still- and moving-picture photography hobbyists. But the crossover of technical hobbyists also extended to dissimilar technologies. In spite of the contrast of the audio electronics of radio with the optical mechanics of photography, a full one-third of respondents to a 1957 ham magazine survey described themselves as hobby photographers, too. People active in multiple hobbies carried technical and social know-how between hobby groups. When Hiram Percy Maxim established the

Figure 1.1
The Amateur Cinema League's magazine ran an article in 1943 suggesting that hobbies such as model building provided good subject material for the hobby of filmmaking. Photograph by Harold M. Lambert, *Movie Makers*, March 1943, page 94, reprinted with the permission of Lambert Studios.

Amateur Cinema League in 1926, he modeled it on the American Radio Relay League, which he had founded previously for ham radio. Maxim created a monthly magazine for members from the club's start, emphasized the importance of national organization, and held contests to increase participation. Half a century later, Wayne Green—long a key player in ham radio publishing—started *Byte*, the first magazine for computer hobbyists. The debut *Byte* editorial alluded to the connection between various technical hobbies by declaring that computer hobbyists had "an emotional kinship with the people who take part in the automotive hobbies."[7]

Model rocket builders, ham radio operators, computer hackers, amateur pilots, and other technical hobbyists shared an inclination to technology despite the differences in the apparatus and methods of their hobbies. All broadly enjoyed technicality—technical devices, technical interactivity, and status in separate technical communities—to the point of envisioning "technical" as a personal trait. With respect to the psychoanalytic phenomenon of identification, I use the phrase that technical hobbyists "identified with technology" to indicate that they reflected upon and represented aspects of the self in relationship to technology. This was an identification articulated through reference to technology more than an identification with the substance of technology. Sometimes a simple statement of brand loyalty, like "I'm a Mac person," inadvertently expresses literal identification with technology. For dedicated technical hobbyists, technical identification formed a fundamental component of personality with far-reaching implications.

A number of scholars have suggested that technology contributes to shaping personal identity.[8] Yet explicit, extensive analysis along these lines hardly exists outside of Sherry Turkle's groundbreaking work on computers. Turkle described personal computers of the early 1980s (the period when they began to find a larger audience but essentially still were unfamiliar devices) as "evocative objects" that inspired users to think about themselves in new ways. The possibilities for creative application of information technology expanded when user-friendly interfaces replaced outwardly technical ones. Interactions with computers then altered basic ideas about human identity. Questions such as the meaning of self for an individual who represented herself differently across multiple online communities spread beyond the group of users and forced a reevaluation of the general concept of "identity in the age of the Internet."[9]

Technical identification takes place within a framework that defines the accepted meanings, uses, and values for technologies. To clarify that this framework is merely the subset of cultural norms specific to technology, I refer to it as "technical culture." As does technical identity, technical culture puts a name to an idea beginning to emerge in technology studies. Lisa Gitelman documented the shifting technical culture—and its relationship to the shifting print culture—that accompanied the phonograph, a process that was guided by manufacturers' gradual education of the public through exhibitions and product labels yet ultimately was commandeered by consumers who saw different uses for the phonograph.[10] Earlier studies that highlighted the relevance of cultural understandings of technology include David Nye's look at the "social meanings" of electrification and Carolyn Marvin's attention to what people were "thinking about electric communication in the late nineteenth century."[11] With reference to specific technologies, a loose concept of technical culture has been circulating for decades in talk of the car culture of Los Angeles, for instance, or the "netiquette" that governs behavior on the Internet.

Technical culture establishes a technology's identity—the perception of what a technology is and how it should be used. Commonplace notions of a technology exist alongside of and contribute to its formal definition. In this vein, Gitelman reminds us that patents, "the official textual identities of technology," implicitly contain "deep-seated assumptions about technological knowledge." But there may be disagreement about a technology's identity, especially soon after its debut. When describing the process by which initial uncertainty about the "meanings or functions" of new media is resolved into mutual understanding, Lisa Gitelman and Geoffrey Pingree suggest, "we might say that new media, when they first emerge, pass through a phase of identity crisis."[12] Persistent opposing views on a technology make technical cultures conspicuous.

Technical hobbyists formed technical identities in two senses. They personally identified with technology and they created identities for technologies. The double meaning evoked by "technical identification" points out that the technical identities of people and technologies are coproduced.

Adopting a particular technical identity can produce social categorizations, just as ethnic, class, religious, and national identities produce social categorizations. The technical culture developed by radio hobbyists, for instance, united hams as a group at the same time that it distinguished them from non-hams. In this way, pursuing a technical hobby led to membership in a separate, technically defined community.

The social separation of technical hobbyists was chosen freely, then reinforced by outsiders' teasing. The popular media regularly mocked hobbyists of all sorts.[13] In the case of technical hobbies, this frequently took the form of unflattering comparisons made to a stereotypical crazed, isolated inventor. The author of *Scientific American*'s monthly column "The Amateur Scientist" during the 1950s described "compulsive tinkerers" as individuals "reclusive by nature" who "grow even more reclusive for fear of being thought mad by non-tinkerers." And Groucho Marx, in a 1955 episode of *You Bet Your Life* that featured a ham radio operator as a contestant, drew a laugh with his jab at "the guy next door who builds things all night in his garage."[14] Non-hobbyists spoke of pleasure taken with machines as an obstacle to human companionship. Some of the most strongly stated critiques came from wives of hobbyists active in amateur radio, high-fidelity music listening, and recreational computing, who referred to themselves as "radio widows," "hi-fi widows," and "computer widows" to encapsulate their feelings of having "lost" their husbands to a hobby technology.[15] Remarks about lone tinkerers seem partly based on the misperception that technicality must dampen sociality. Quite the contrary, technical hobbies welcomed participants into lively social circles.

Thousands of clubs formalized technical hobby communities. Hobbyists spoke of clubs, especially those linked through national organizations, as structures that could help to legitimize their activities, to clarify what the hobby was, and to increase its popularity. The editors of *Modelmaker* magazine urged readers in 1924 to join their local chapter of the American Model Engineer Society, suggesting that membership would give the hobby "a boost." "The more the Societies develop," they figured, "the more interest will be taken in Model Making in the U.S." In 1942, *The Model Aircraft Handbook* recommended membership in the Academy of Model Aeronautics, which it called "the official governing body of model aviation in America," to facilitate comparison of records and participation in competitions and as a strategy for "attracting additional members."[16] Enlarging clubs improved networks for socializing with like-minded individuals and for pooling expertise.

Supplementing the practical lessons hobbyists acquired through clubs and informal social-technical networks was a vast body of hobby publications. Handbooks, magazines, and club newsletters carried a mix of detailed how-to guidance and you-can-do-it cheerleading, the proportion differing over time and according to editorial slant. In the case of hobby photography, when the technology was new, amateurs and professionals alike faced daunting challenges. The editor of the 1911 *Encyclopaedia of Early Photography* required nearly six hundred pages describing chemicals, processes, and equipment just to produce what he "intended essentially as a simple guide to photographic process." Technical hints and articles filled issues of both *American Amateur Photographer* and *Popular Photography* when they began in 1889 and 1912, respectively, making them read like technical manuals released in monthly installments. In the 1930s, when hobby photography required fewer technical skills, *Home Photographer and Snapshots* concentrated instructional material in a "Beginners' Section," "For the help and encouragement of those, old and young, who are keen enough on photography to wish to make more and better pictures."[17] Varying levels of technical content then served to differentiate hobby publications, with other magazines continuing to meet the demands of technically inclined readers into the age of point-and-shoot photography.

Technical hobby communities encouraged hands-on activity and celebrated the virtues of learning by doing. The same rhetoric had filled industrial arts textbooks since the start of the twentieth century. Typical of this style, *Home Handicraft For Boys: Learning Through Doing* (1935) praised "the average wide-awake boy, with true Yankee spirit, [who] goes ahead and finds out what he can do." The book suggested that boys use their free time for "building radio equipment, constructing toys, devising things for mother and the home, making repairs about the house," all because "We learn most quickly through experience."[18] Industrial arts curricula that incorporated technical hobbies endorsed the educational value of leisure tinkering. Plans for building radios appeared in textbooks from the 1920s, and a 1942 text advocated teaching aviation hobbies to help "American youth to become air-minded."[19] Outside of the classroom, promoters of technical hobbies emphasized that these forms of recreation had the benefit of being "instructive as well as constructive." The editors of *Modelmaker* magazine wrote that assembling models offered "a fuller understanding of the principles governing the successful operation of a piece of mechanism."[20]

For technical hobbyists, hands-on activity represented more than a method for acquiring information, and in fact was a guiding tenet of their technical practices. Hobbyists of all sorts were by definition busy, engaged, and productive. Taking that approach to technology marked an important difference between technical hobbyists and users of technology. As early as 1890, hobby photography magazines warned participants that turning over printing procedures to a commercial finisher would result in the loss of "a good part of the enjoyment there is in the pastime." Other technical hobby publications bluntly stated that reduced technical activity would jeopardize status as a hobbyist. A chapter on "Model Airplane Club Organizations" in a 1928 handbook recommended that "Only boys who are definitely planning on constructing models should be admitted, as others will harm rather than help the group." Non-builders so severely violated the community culture that the author declared, "Definite procedure for dropping dead timber should be adopted and the rules strictly enforced."[21]

Technical hobby communities considered the extent of members' interactivity with apparatus to be a measure of personal commitment. With the availability of snapshot cameras, hobby photography groups distinguished "snapshooters," who just pushed a button with little technical knowledge or interest, from "serious amateurs," who exhibited great technical skill. Of course, many leisure photographers fell in between these extremes. The division into casual and devoted users of a technology functioned as a moral categorization imposed by a hobby's technical culture. Some model airplane clubs labeled members as either ground men, flying cadets, pilots, aces, or assistant instructors. To climb the ranks, a hobbyist had to "pass an examination on some phase of aviation as well as to construct certain models with set specifications and requirements."[22] These kinds of hierarchies made explicit how technical hobby communities based social standing on technical expertise and accomplishment.

To remain active, hobbyists required technologies that were not blackboxed. The term "black box" refers to a device where only the inputs and outputs are apparent and the functional mechanism is unseen and unknown by the user. (In the context of consumer technologies, devices with these properties often are described as "user-friendly." The mildly negative connotation of "black-boxed" better fits the perspective of hobbyists who took an interest in technology.) When trying to sell devices to the largest possible market, manufactures tend to reduce the knowledge needed by users, creating simplified products that approach black boxes. Hobbyists sometimes adopted a technology for leisure use before its operation had been streamlined for mass consumption. In other situations, hobbyists avoided the constraints of ready-made equipment by building their own or, even more commonly, modifying purchased gear to individual tastes or purposes. When choosing from a range of available consumer technologies, hobbyists gravitated toward the more interactive apparatuses. Amateur filmmakers, for example, reacted to the creation of an accessible version of the hobby's central equipment by selecting quasi-professional equipment. As Kodak and Bell and Howell diversified their product lines to include easy-to-use and more affordable movie cameras in the 1950s, serious amateur filmmakers differentiated themselves from camera owners who casually recorded children's birthday parties by buying the more complicated, expensive cameras that allowed those with skill to retain technical control. A concern that hobbyists would become indistinguishable from other consumers in part prompted audiophiles in the 1980s to speak out against digital compact disc technology, which threatened to democratize the market for high-quality audio equipment.[23]

Hobbyists striving to display technical proficiency or dedication risked obscuring the recreational component of their activities. Throughout the twentieth century, social norms dictated that middle class Americans should engage in some type of leisure, understood in contrast to work.[24] To meet this expectation, hobbyists who aspired to reach professionals' levels of skill and deep involvement with technology emphasized that they were not exactly replicating work activities during free time. "A direct appeal to [Amateur Cinema] League members to undertake filming that will be artistically significant" in that club's magazine in 1928 implored amateurs to "try to get as far away as possible from the professional in subject matter and as close to him as possible in workmanlike technique." Three decades later, analogous instructions appeared in the hobby literature of "sound hunters." Participants were told to use their recording devices to capture everyday sounds, with a "fresh approach," "preferably not in the style of professional radio broadcasting."[25] These minor distinctions hardly opposed professional methods.

In some instances, hobbyists more strongly broke from industrial technical culture. Marc Perlman observed that two groups of technical hobbyists, audiophiles and those who continue to operate the long-outdated TRS-80 model of personal computer, both "defend a moral order privileging activity and independence." These hobbyists create independent technical cultures by partially rejecting professional technical culture. With audiophiles, this is manifested in the appeal to personal listening experience over scientific analysis by audio engineers. Users of the obsolete TRS-80 express a dissatisfaction with contemporary personal computers.[26] Some technical professionals spoke of pursuing closely related hobbies in their spare time in order to break from workplace practices. A member of the Homebrew Computer Club described this early computer hobby group as "a bunch of escapees, at least temporary escapees from industry," who appreciated that "the bosses weren't watching" and who "knew this was our chance to do something the way we thought it should be done."[27]

To separate their tinkering from commercial or professional pursuits, hobbyists invoked the language of "amateurs." But the amateur-professional distinction put forth by hobbyists was problematic at best and in many cases spurious. Not all hobbyists met the commonplace requirements of amateurs, who pursue an activity only for the love of it and not for profit. In defining "amateur photographer," *The Encyclopaedia of Early Photography* claimed that "a little payment" should not move a hobbyist from the category of amateur to professional.[28] The Amateur Cinema League also was concerned more with "the spirit of an undertaking" when defining "amateur" than with "hair-splitting" on the matter of income. Hobby publications in general after World War II frequently contradicted amateur principles by suggesting how to derive a profit from hobbies.[29] Further confusing the issue of technical hobbyists' amateur status was the fact that many held technical jobs, sometimes jobs very similar to their hobbies.[30]

Affiliation with technology professionals did prove useful at times, and hobbyists then exploited the ambiguity of their position. For one, it allowed practitioners to make the justification that technical hobbies were "more than a pastime." The editors of *Modelmaker* magazine suggested that assembling models should instead be thought of as "the Science of Engineering in miniature as it includes all the essential features of its larger prototype." This contributed to another positive aspect of technical hobbies' connection with the working world: expertise developed in basement workshops could be parlayed into technical careers. Though not every hobbyist would become a captain of industry—as had been the case in a few oft-repeated stories—learning by doing injected vital skills into the workforce. U.S. Naval Commander and flight instructor Richard E. Byrd, in his introduction to a 1928 model airplane handbook, predicted that "From their [model aviators'] ranks will come the designers of the new ships of the sky that will supersede present aircraft." Time proved him correct. *The Model Aircraft Handbook* published during World War II bragged that "The aviation industry, the Army Air Corps, and all the other branches of full-scale aviation have drawn heavily on our hobby for new fliers, designers, and workmen." Based on this leisure-to-career trajectory, aviation hobbyists insisted that, at a time "when 'Keep 'Em Flying' is a keynote in national defense," aviation recreation would provide critical inspiration "to 'Start 'Em Flying.'"[31] Equivalent testimonials to the instruction and career enhancement available through tinkering appeared across all technical hobbies, and engineers and scientists throughout the twentieth century regularly credited boyhood technical hobbies with having sparked lifelong interests.

There were many personal rewards from pursuing technical hobbies. Participants gained skills and a sense of accomplishment, found fellowship in a distinct community, claimed a place in the formidable technical world, reached new self-understandings, and improved career opportunities, all in an enjoyable recreational context. The experiences of hobbyists with technology additionally had much broader effect. As they formulated technical identities, technical hobbyists influenced thinking about technology beyond hobby communities. The meanings and uses hobbyists arrived at for technology and their realization of implications the technology had for self perception caused non-hobbyists to question their assumptions about technology. Hobbyists often were enthusiastic early adopters of new technologies, in which cases their role as leaders of change in technical culture was quite clear. Their more subtle contribution also was important as a technology aged. The existence of a productive, leisure form of a technology after a consumer or industrial form of that technology had become standard outside of the hobby context enriched the general technical culture by posing a vital counterpoint.[32]

New audio, visual, and writing technologies reconstituted human experience at the deep levels of perception and comprehension. The evidence Friedrich Kittler drew from Irmgard Keun's *Rayon Girl* (1932) in support of this assertion is instructive here. The title character, Doris—inspired by listening to the radio and hearing a neighbor's gramophone—imagines that a screenplay, rather than a poem or novel, is the literary form best suited to her life story. "I want to write like a movie, because that's the way my life is and it will soon be more so," she declares. "And when I read it later, it will be like a movie—I will see myself in images."[33] Amateur filmmakers did not need to write movie-like autobiographies; they filmed autobiographical movies. Technical hobbies offered opportunities for working through altered visions of self and surroundings brought to light by technological change.[34]

Consider what happened as cameras entered novice hands. Historian Robert Mensel situates the beginning of amateur photography in the context of the late-Victorian "weightlessness" resulting from "the accelerating pace of urbanization, secularization, industrialization, and scientific discovery." The background that Mensel calls a "sense of social and psychological dislocation among bourgeois Americans" I would foreground and suggest was part of the appeal of hobby photography.[35] Individuals who felt dissociated could reestablish a connection to a world increasingly dominated by technology by picking up and mastering one of the modern machines. Snapshooters and serious amateurs alike moved beyond the experience of being photographed and explored what it felt like to be the photographer, to frame the world in a "viewfinder," to record visual information. They grappled with a new form of sight, refracted through the camera and fixed on paper. In 1888, George Eastman gave one of the first Kodak cameras to Henry Strong. Strong was familiar with photography: as Eastman's business partner and financial backer, he had a substantial financial investment in popular photography. Even so, Strong had an epiphany when he used the camera himself on a cross-country trip. "He was tickled with it as a boy over a top," Eastman reported in a letter to a friend. "He apparently had never realized that it was a possible thing to take pictures himself."[36] Releasing the shutter with his own hand, Strong viewed the camera, and in some sense, the world differently. He became personally invested in popular photography.

The press frequently reported, with a mixture of enthusiasm and fear, that the debut of a technology promised to wholly reshape everyday life. To thrive in the dawning technical age, therefore, would require an adaptation greater than merely learning to use some new device. Those who joined in predicting revolution called for an extensive process of enculturation to the arriving technical culture. One extraordinary example of this line of argument recommended that schools introduce a curriculum appropriate to the culture of the airplane. In response to their belief that students were "living in a new world—indeed, they are in a new age," members of the Aviation Education Research Groups active at the Teachers Colleges of Columbia University and the University of Nebraska created a comprehensive series of textbooks. The Air-Age Education Series, produced in cooperation with the Civil Aeronautics Administration, related various subjects to aviation. Among the titles were *Human Geography in the Air Age, Social Studies for the Air Age,* and *The Biology of Flight.*[37] The specific proposals these texts made for how to prepare for a new technical culture, though, were atypical among vague yet urgent warnings that sweeping change was coming.

Taking technology into their own hands, technical hobbyists were ready for—often leaders of—revolutions in technical culture. Newspapers and magazines of the 1920s and 1930s acknowledged this when they elided model building with actual flying in a general celebration of flight. Hobbyists then appeared to stand at the vanguard of the forecasted airplane culture. In his cultural history of the "technological enthusiasm" for aviation, Joseph Corn explains that hobby activity gained support from "airminded adults [who] believed that the boys and girls who were building and flying model airplanes were irrefutable proof that the prophesied air age would, in time, be realized." "The image of youngster and model airplane" then functioned "as a kind of icon" for aviation enthusiasts.[38] Together these strains of thought produced a circular logic that boosted both aviation culture and technical hobbies. People who saw on the horizon a world inflected everywhere by aviation encouraged children to construct model

Figure 1.2
A row of men waited to watch boys race model airplanes in Los Angeles. Photograph from *Aerial Age Weekly*, 16 February 1920, page 669.

airplanes and fly them in competitions during their spare time. Participation in recreational aviation, in turn, was interpreted as evidence of the spread of aviation culture (see figure 1.2).

The coproduction of the technical identities of people and technologies fueled the evolution of technical culture. For hobbyists, this process was direct, intimate, and profound. Hobbyists encountering apparatus first-hand consciously reevaluated their ideas about technology. And the experience of existing in a social group segregated by recreational use of technology encouraged hobbyists to identify a technical component within themselves. Those not immediately using technologies produced technical identities through a process that differed by degree rather than kind from that of hobbyists. Back to Keun's 1932 character Doris: she was surrounded by a culture that valued movies as an entertainment technology. She went to see movies and in the theater absorbed moving images. The experience of watching movies became part of how Doris understood the world. She may not have gone so far as to call herself "technical," but Doris thought with movie technology; she saw herself in movie-like images. This shift in perception further enhanced Doris's sense of living in a movie culture.

Hobby applications of technology that fell outside of mainstream technical culture gave rise to explicit debates about the proper role and place of technologies. At the turn of the twentieth century, the press depicted hobby photographers as strange characters who possessed dangerous equipment. Photography magazines joined general publications in calling early amateur photographers "camera fiends" and using similar language that associated hobbyists and their cameras with violent and treacherous misdeeds. The sharp division of public opinion over the portable camera—a drama whose episodes included the proposal of a federal law to limit exhibition of photographs, the arrest of a photographer for selling manipulated images, and a court injunction against a magazine for printing an actor's photo without his permission—contributed to the development of legislation to protect privacy. Arguments regarding technical hobbies only rarely reached the courtroom. Local zoning boards, however, heard neighborhood disputes about the odd-looking antennas ham radio operators installed on their properties. And the spousal bickering provoked by high-fidelity audio hobbies gained public attention as the topic of humorous articles in the hobby and popular press.[39]

The balance of this book focuses on radio communication as a technical hobby in mid twentieth century America. Hams did something unusual by adamantly continuing to tinker with radios long after the standardization of commercial broadcast radio. Their pastime occasionally drew snickers and raised suspicions. In response, radio hobbyists portrayed their activities as exercising skills critical to the electronics industry and to national defense. The negotiation of a hi-tech identity adapted to the Cold War household distinguished a community of men with a unique perspective on technology. The historical value of this case study stems mainly from its singularity. While certain conclusions can be generalized to technical hobbies, the greater lessons for contemporary society come from witnessing the contrast of amateur radio with other mid century technical cultures.

Whether serving as leaders or provocateurs, hobbyists demonstrated diverse options for technical culture. Hobbyists engaged with technology in a way that was fun, collaborative, educational, intense, and creative. These methods and values were independent from, and at times in direct conflict with, the technical culture of profit-driven production. In 1976, Bill Gates issued an "Open Letter to Hobbyists" in an overt attempt to transform the culture of computing. Electronics hobbyists tinkering with the first personal computers were exchanging software as freely as they always had exchanged ideas and technical manuals. From the standpoint of Micro-Soft (as it was known then), which had spent tens of thousands of dollars to create a version of BASIC for the Altair computer, sharing copies of the program was "theft" and would "prevent good software from being written." "Who can afford to do professional work for nothing?" Gates asked rhetorically in the letter.[40] But that was precisely what technical hobbyists had been doing for decades, in part by pooling their efforts. Over the next quarter century, the culture of personal computing changed considerably without entirely suppressing the impulse to share. A cooperative spirit persists today in open-source development, in the legitimate distribution of free software, and in cavalier attitudes toward the illegitimate copying of proprietary software. This spirit is a legacy of hobbyists and a reminder that there exist alternative ways of using and relating to technology.

2 The Culture of Ham Radio

For a pastime, two-way radio was highly regimented. The federal government regulated hobbyists' use of the airwaves, and hams extended the state's control by devising a protocol for personal behavior and relationships and for the style and content of communications. In return for accepting the ham radio culture, participants gained a sense of belonging. Outsiders recognized this bond when they observed that radio hobbyists were "clannish" and "a closely knit clique."[1] Hams formed a community through the same general practices of other social groups. They set conditions for membership, established rules of conduct, taught values, and developed a specialized vocabulary known only to insiders. What made hams' culture different was its basis in technology. The norms of ham radio hold the key to understanding the role technology played in creating community and the process of a community making a technology its own.

Codes of Behavior

Learning the group culture was essential to becoming a ham, and ham radio publications taught behavioral expectations to new hobbyists right along with technical lessons. The *ABC's of Ham Radio* welcomed readers to "the ranks of the grandest hobby in the world—the great international fraternity of radio hams!" then indicated in the very next sentence that "To really belong, you're going to have to go along with the standard operating procedures universally accepted by radio amateurs." Most manuals devoted a chapter to operating a wireless station, including an overview of on-air etiquette. One author noted that "a sense of courtesy is important" and told hams not to transmit on frequencies already in use. With surprising regularity, handbooks also endorsed general personal "qualities of the true amateur" such as "inquisitiveness, persistence, improvisation, imagination and an open mind." The exchange of technical ideas through magazine columns was cited on one occasion as a testament to the fact that "The amateur spirit has always been characterized by friendliness, helpfulness and an eagerness to share one's knowledge, tricks and pet circuits with others."[2] The constant stream of brief prescriptions of norms and values in hobby publications served as a powerful source of enculturation into the ham community.

A concise, and the best known, list of good hobbyist conduct was the "Amateur's Code" distributed by the American Radio Relay League (ARRL). "The amateur" portrayed there is "gentlemanly," "loyal," "progressive," "friendly," "balanced," and "patriotic." The League has printed these six traits prominently in the front of its annual *Radio Amateur's Handbook* since the 1920s. Underscoring the instructional nature of the code, a didactic explanation followed each adjective. A ham's progressivism, for instance, meant that "He keeps his station abreast of science. It is built well and efficiently. His operating practice is clean and regular." The League's role as a lobbying agency shone through in deeming a hobbyist "gentlemanly" for abiding "by the pledges given by the ARRL in his behalf to the public and the Government."[3] The ARRL's "Amateur's Code" provided a model for hams to live up to and presented a favorable image of hams to outsiders. Given how frequently the popular press reprinted the standards as if they offered a neutral description of hobbyists, the "Amateur's Code" succeeded as a form of public relations.

The social ties of the ham community exerted peer pressure to enforce the rules set for members' behavior. Praising the effectiveness of "self policing" within hobby radio, a *CQ* magazine article called "The weight and influence of amateur approval [. . .] a very strong element in forcing the amateur to abide by the rules." A handbook instructed, "At all times keep your conduct beyond reproach," and tried to win compliance by reminding the reader, "You represent the amateur fraternity—any action on your part, good or bad, will reflect on all other hams." When the "fraternity" roster had swelled to more than a quarter million in the United States alone, another manual stressed that the "number of stations in our crowded bands poses a serious threat to our enjoyment of ham radio if we do not all operate courteously and intelligently."[4] Hobbyists who did not meet community expectations were subject to criticism, punishment, and in extreme cases expulsion.

The strategic potential that set wireless communication apart from most hobbies subjected it to a level of state scrutiny unheard of for other leisure activities. The power of the federal government stood behind the only official barrier to entering the ham community: obtaining a license to operate two-way radio. Licensing of ham radio began under the Radio Act of 1912 and varied little over the next eighty years. The Federal Communications Commission (FCC) required prospective hobbyists to demonstrate knowledge of electronics theory and radio regulation in a written exam and the ability to send and receive Morse code in a test performed with wireless apparatus. The FCC contained amateur conversations to particular bands of the radio spectrum, restricted the power of transmitting equipment, required hobbyists to log all contacts, and monitored the airwaves for infractions. Because they regarded state control as a tribute to their strength, hams accepted federal licensing and communication regulations as the first level of hobby radio rules.

In the early 1940s, wireless hobbyists trying to change their image from tinkering pranksters to upstanding citizens volunteered to help the FCC track down unlicensed operators. The American Radio Relay League spoke of lending assistance with enforcement as a tactic to keep hams on good terms with regulators. When the FCC caught a notorious "unlicensed punk" in 1941, the ARRL chided members for not having found him and called for improved "policing" within the hobby. The League reasoned that "our interests require that we show no tolerance either to bootleggers or to violators of the FCC's special orders." Defense of community boundaries further motivated hams to turn in illegal operators. Monthly club bulletins offered a timely format for calling attention to mischievous on-air behavior. The newsletter of the Northern California DX Club, for instance, exposed an operator suspected of using false credentials after confirmation cards a member sent to him had been returned marked "addressee unknown."[5] Joining together in this way to ostracize rule breakers from the on-air community increased solidarity among upstanding wireless operators.

Ham radio licenses functioned as membership cards signaling inclusion in a technically elite club. Like station licenses for commercial radio and television broadcasters, all hobby licenses in the United States began with "W" or "K." On amateur licenses, the initial letter was followed by a numeral—designating which of the nine FCC geographical districts the operator lived in—and two or three additional letters. The alphanumeric "call signs" lent hams legitimacy and, in some cases, reflected the duration of the holder's radio activity. When the FCC first issued amateur licenses, all began with "W" and contained three letters total. The creation of calls that began with "K" and of calls containing four letters only occurred once the number of short "W" calls was exhausted. After the FCC introduced these new calls, a ham with a short "W" call like W3CT could be recognized immediately as a longtime license holder compared to a ham operating under W8JBH or K2MJW. Call signs became hobby community nicknames, and club newsletters frequently referred to members by license number instead of name. Even many outsiders learned to recognize the basic form of FCC licenses, so that a car with a call sign vanity license plate stood out as belonging to a ham radio operator.[6]

Although hobbyists enjoyed being distinguished as more technically adept than average citizens, many objected to the technical hierarchy imposed within their ranks by the FCC's "incentive licensing" program. Beginning in the 1920s, the FCC offered various amateur license grades. Hams who passed an advanced theory test and exhibited faster Morse code sending and receiving skills earned additional operating privileges and bragging rights in the form of "Extra" or "Technician" licenses.[7] An editor at *CQ* magazine in 1966 blamed the internal division of hobbyists according to ability for provoking "fierce in-fighting," and the Commission's expansion of the incentive program a few years later angered hams. Letters of protest poured in to *CQ*, accusing incentive licensing of undermining the "unity" of "the radio fraternity." One writer argued that with "the old days of major electronic breakthroughs by amateurs" a distant memory, it made sense to "bring back the fun of amateur radio" and "junk the snob appeal of incentive licensing." Based on the negative reaction, *CQ* estimated that if "a vote had been taken of *all* licensed amateurs" on whether to expand the incentive licensing program, "it would have been defeated by an almost three to one margin."[8]

One way hams displayed their technical identity was by using Morse code. Their admiration for the code as the ideal form of communication stemmed from the importance granted to coding skills in the FCC licensing examination and from hobbyists' appreciation of how the code transformed language. Tapping out sequences of short and long electrical pulses on a telegraph key required human synergy with machinery and gave words a technical feel. Still, the sender's personality transmitted through the machine. "Code operators quickly learn one another's 'touch,'" wrote an Army radio specialist. "The way a person sends code is almost as distinctive as his voice."[9] Hams referred to this human accent detectable in code transmission as the sender's "fist." In the early days of wireless, Morse code was the only way to transmit a message. Long after it became possible to speak over the airwaves, numerous articles in radio publications and speeches at club meetings extolled the virtues of Morse code. Hobbyists praised the code as reliable and versatile and also called attention to "a special beauty in perfectly sent code and a certain emotional rhythm" to some words. The further claim that Morse code was "a widely understood international language [...] that links hams together throughout the world regardless of their individual, indigenous languages" was a gross—but not uncommon—exaggeration because Morse code encoded the alphabet, not words or concepts.[10]

The code set adept hams apart from confused outsiders. The written "key" that assigned a combination of dots and dashes (representing short and long electrical pulses) to each letter of the alphabet was widely available, but the challenge of applying Morse code kept it somewhat at the level of a cipher. Only with practice and, according to hams, patience, dedication, and attentiveness was it possible to transform thoughts fluidly into tapped electrical pulses or to hear phrases emerge from patterns of short and long tones (see figure 2.1). Communicating by Morse code created privacy in public. Tales of getting a fellow ham's attention across a crowded room by speaking his call sign in Morse code—substituting the syllable "dit" for each short pulse and "dah" for each long pulse—were frequently and fondly recalled. One hobbyist described secret exchanges he had with his brother while double-dating as teenagers, Morse code giving them the freedom to discuss "the characteristics of our dates in their presence without their knowing it!"[11]

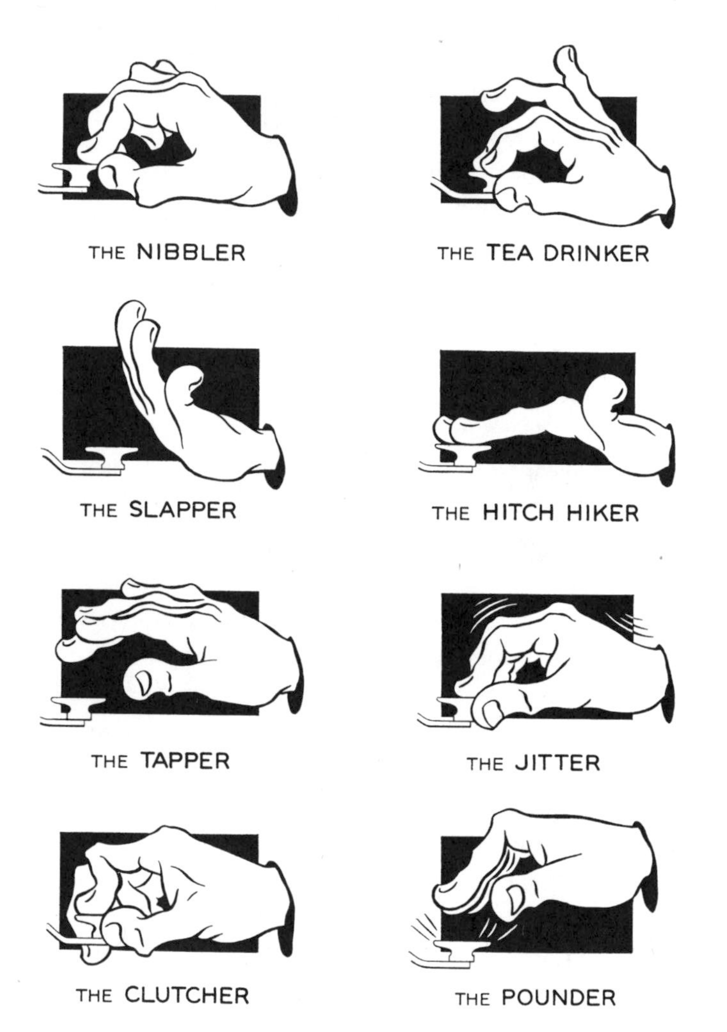

Figure 2.1
An illustration in an article on "Proper Sending Techniques" caricatured several undesirable styles for tapping out Morse code on a telegraph key. From *MARS Bulletin*, March 1952, page 30.

The main alternative to wireless communication by Morse code was voice or "phone" operations. In this case, hams with the proper equipment could just speak. Phone transmitters fell within financial reach of the average hobbyist after World War II. Surveys about operating habits conducted by radio magazines found that the typical postwar ham split his time between coded and spoken operations, spending about twice as much time using phone as code. A small portion of hams, about 5% in 1957, worked only in code.[12] The simplicity of voice operations led to continuous, passionate debates that code better suited a technical hobby. When the FCC dropped knowledge of Morse code from the requirements for a basic amateur radio license in 1991, fierce opposition to "no code" licensing included a "know code" movement among Morse-loyal hams who insisted that the code remained vital to modern operations.[13]

Preference for code over spoken communication reflected a desire to rationalize language. Transmitting by Morse code processed words through technical apparatus and removed the voice from communication. Advocates of the code claimed that translation into its binary system of electrical pulses eliminated vagueness. "Communicating by dot and dash," Howard Pyle contended, was "far more accurate than the spoken word." Since the complexity of Morse operation made it unlikely that the code would be perfectly sent and received, and in light of hams' acknowledgment that the "fist" inflected Morse with the sender's personality, arguments that the code assured clarity sounded like appeals to its pure technicality. Attempts in hobby publications to establish a beneficial association of hams with the military by pointing out that both used the code only made the code seem more disciplined. With encrypted, systematized language, hams also reduced the risk that radio conversations would be associated with what they saw as women's idle chatter. The explanation by a female hobbyist in 1948 of her preference for Morse code suggests the existence of a gendered spectrum of wireless communication with women talking as the most feminine mode, men coding the most masculine, and men talking and women coding falling somewhere in between. The intrusion of what she called "too many $%&'( )* unlicensed wimmin [*sic*] (wives, gal friends, etc.) cluttering up the phone bans with chin music," led Carol Witte to conclude, "any self-respectin' licensed gal wouldn't be caught dead blabbin' fer [*sic*] hours on a mike—nor a good OM [male] operator, either."[14]

Morse loyalists battled phone loyalists for territory on the airwaves. It is difficult to document these feuds, which usually were limited to a heated exchange of words, but a few escalated to the point that regulators became involved and left a paper trail. The FCC counted Myron Premus among the "considerable number of amateurs in the Buffalo and up-state New York area" who fought to eliminate code operation from portions of the radio band in the early 1950s. After receiving "complaints regarding the manner in which he has operated his radio station," the Commission evaluated whether to renew Premus's license. The subsequent investigation found that Premus had "caused willful interference" to hams using Morse code by making "one-way communications consisting of disparaging remarks either about the operator or his manner of operations."[15] Opposition by Premus and others to Morse code may have disturbed hams' conversations, but it did not threaten ham identity.

The hobby radio community made language its own and clarified group membership by adopting jargon and abbreviations known only to insiders. In a few cases, jargon arose from the desire to convey non-words through Morse code, such as when hams indicated laughter or sarcasm by signaling "hi hi." Hobbyists used abbreviations to shorten Morse code transmissions and carried these into their regular writing. Substituting "vy fb" for "excellent" in a hobby publication reduced keystrokes. The symbolic efficiency of abbreviations further supported hobbyists' portrayal of radios as efficient devices and radio operators as efficient people. Even more significant, the abbreviation lent the text a bit of technicality by associating it with Morse code. Many of the abbreviations used by hams came from a system devised by telegrapher Walter P. Phillips in 1879. Hobbyists also took up telegraphers' "Q signals," three-letter combinations beginning with the letter "Q" that represented common phrases. "QTH" served as a quick way to ask a station location, for example, and even functioned across language barriers.[16] Only the hobby community expected members to be fully conversant in jargon, with the FCC licensing examination merely testing the essential Q signals.

When hams peppered spoken and written language with abbreviations intended for efficient Morse code transmission, they gave all forms of group communication the flavor of ham radio. This propagation of hobby culture accounts for the persistence of awkward habits such as interrupting the flow of conversation with another ham by saying "hi hi" instead of simply laughing. A few sticklers insisted that the Phillips code and Q signals could be used "properly" only within the Morse system. During phone conversations, in person, or in print, this minority said, it was "more natural" to just say or write the complete phrase rather than the abbreviation. In response to "several years" of what it called "weak and withering attacks against that traditional amateur workhorse: The Q-Signal" by those who favored normal, full words, *CQ* magazine defended spoken codes as more than a linguistic convenience. The Q signals, according to the editorial, "catch the imagination of the newcomer" and formed part of "amateur radio's character." In the late 1960s, "the radio amateur's most individualistic jargon" also helped separate hams from Citizens' Band hobbyists, who the *CQ* editor described as using "mundane and lackluster phrases."[17] Asking "What's your QTH?" instead of "Where are you located?" indirectly inserted Morse code into plain English, signified membership in the ham community, and left outsiders scratching their heads.

Hobbyists valued clear, standardized speaking during phone operations. They gave the practical justification that distant communicators had trouble understanding each other's accents, especially when reception was poor. Extreme language regimentation appeared to represent an attempt to strip away the individuality of human speech and replace it with a mechanical uniformity. Annoyed with "hams who abuse the ears of their listeners," Don Fox wrote a guide to help hobbyists determine whether they suffered from "mumble-itis." Fox described ham radio as focused on "getting a thought to somebody else by way of intelligently combined sounds." He harped on "proper enunciation" and directed mumblers to "books on the subject of proper speech and the training of the speaking voice."[18] While calls for such broad corrections of speaking style were rare, all hobbyists agreed on the need for linguistic precision in certain situations.

Hams coped with the similar-sounding names of letters of the alphabet—crucial for conveying call signs—by associating distinctive words to each letter. "KB3DF" would read out his call as "kilowatt bravo three delta foxtrot," for example. Several supposedly "standard" phonetic systems circulated among hobbyists, with none dominant and each freely varied in application. KB3DF's preferred rendering of his call broke from the International Civil Aviation Organization's phonetic list only in substituting "kilowatt" for "kilo." This particular customization of an outside template to the hobby was quite common and related to the special meaning that a kilowatt held in ham radio as the maximum legal operating power. Disdainful of other "cute" alphabet-word pairings that "have no business being used on the air," an ARRL handbook reminded readers that "there is a definite advantage in using a standard phonetic alphabet."[19]

Speaking habits, transmitting practices, and even the content of radio exchanges were disciplined through surveillance. The FCC monitored the airwaves mainly for operating violations. In 1946, *CQ* compelled readers to obey regulations with the threat that the Commission's "mobile units are continually patrolling the country, stopping in cities to observe local activities, and listening from vantage points for unlicensed stations." Hobbyists meanwhile handled the policing of the community's internal communication rules. If they did not like what they heard in the course of scanning the amateur band, hams freely critiqued operators and occasionally passed matters on to federal authorities. It was the verbal reprimands Myron Premus had issued to fellow hams, for instance, that prompted his investigation by the FCC. When Premus "noticed off-frequency operation, over-modulation, or other operations not in accordance with the Commission's rules," he called the offenders "lid," "louse," "jerk," and "hollow head." One ham found Premus out of line for using such language on the air and alerted the FCC. In defense of Premus, other hobbyists expressed their own frustration with the "many dopes on that band that should not be on." They sympathized that "We cannot take away their licenses" and that derisive name calling was the strongest punishment that could be meted out by the ham community. The FCC agreed with the assessment that Premus had been incited to speak out, though its report cited improper operating procedures as the provocation rather than a breach of hobby standards.[20]

A gentlemen's agreement protected wireless discussions exposed to all ears. Claiming that those who only listened to the radio lacked the discretion of two-way radio operators, a tale in *CQ* magazine directly linked the attributes of a technology with the character of its users. The author described his teenaged neighbor as fascinated by what hams revealed to anyone who might tune in with a shortwave receiver. On meeting a ham in person, the shortwave listener repeated embarrassing personal information he had heard disclosed over the air. To stop this impolite behavior, the author helped the teenager study for a ham license because "no ham dares tell what he knows about another." The community believed that two-way communication made hobbyists discreet through a control mechanism absent from shortwave listening. What kept hams from gossiping was the risk of retaliation, the fact that "the other knows as much about him."[21]

State control of the airwaves further disciplined radio operators by effectively squelching political conversations. Hams recognized they were "involved with, formed by, and regulated by politics." Yet fear that ideological battles would result in tighter regulation by the federal government led hobbyists to pragmatically refrain from political activity "unless it is something for the good of amateur radio," stipulated a 1935 club bulletin, "and then, only when it is absolutely necessary."[22] The ARRL hired professionals to lobby for radio rights, and many smaller organizations and individuals spoke with their representatives in Washington whenever competing forms of communication encroached upon amateur bands or when international tensions threatened to silence the hobby. Otherwise, ham radio culture dictated that there was to be no discussion of politics on the airwaves, at club meetings, or in hobby publications.

Hobbyists connected their apolitical stance to radio technology, offering the logic that the rational, scientific character of wireless communication demanded politically neutral operators. In 1961, a ham alerted members of his radio club to what he considered "SICK broadcasts" announcing the formation of the Anti-Communist Amateur Radio Network. "Amateur radio should be held aloof from these things," he insisted in the club newsletter. "There is no place in amateur radio for these groups be they religious, anti-communist, pro-African, or what have you."[23] Hobbyists so strongly opposed the spreading of political messages via radio that they occasionally broke the law to enforce this community rule. The Student Information Network used the amateur bands to coordinate strikes on college campuses in May 1970 in response to the Ohio National Guard having killed four students at Kent State University during a protest of the decision to send U.S. troops into Cambodia. Some hams upset by this disruption of "the peace and sanctitude" of the airwaves wrote letters to hobby publications. Others took more direct action, orchestrating "widespread" jamming of the Network's communications. Though *CQ* magazine supported the ham community position that "amateur radio should not be used as a sounding board for politics," it pointed out that "no such limitation is legally imposed on us." The only law breakers in this conflict were those who willfully interfered with the Student Information Network.[24]

The neutralization of on-air language and topics had drawbacks. That a hobby formed around communication had "by and large pretty dull things" to communicate weighed on some hams. Sporadic editorials and articles lamented that "Most QSO's [contacts] are a crashing bore." To solve this perceived problem, the editor of one ham magazine solicited "an article or two which would give all of us some good hints on how to plunge into a conversation with some chap we've never met."[25] Guidelines for discussions commonly included asking about a ham's other hobbies and his location and avoiding talking about the weather unless it was severe.[26] The ARRL encouraged extended on-air conversations by designating an award for "rag chewers." Any ham who reported "a fraternal-type contact with another amateur lasting a half hour or longer" qualified for membership in the Rag Chewers Club. The award rules specified that this could not include time spent chatting about technical aspects of radio. Rather, it was to be "a solid half hour of pleasant 'visiting' with another amateur discussing subjects of mutual interest and getting to know each other."[27] A certificate from the Rag Chewers Club offered a lighthearted incentive that matched the lighthearted exchanges the ARRL sought to promote. By comparison, there was a certain irony in cases where the ham community attempted to rigidly discipline conversations by criticizing hobbyists for not being sufficiently casual.[28]

The hobby community fostered a particular kind of sociability by endorsing selected forms and styles of off-air communication. The first non-radio contact between two hams usually was the exchange of postcards called "QSLs." ("QSL" is a Q signal for "I acknowledge receipt.") Through these cards, ethereal, fleeting, auditory conversations took on a material, enduring, visual reality. It was common for a ham to customize his confirmation cards with images and text that conveyed something about himself, his locale, or his relationship to the hobby and to create a card "truly represen-

Figure 2.2
The confirmation postcard of W8CPC, a ham from Buffalo, New York, carried a drawing of a buffalo and the initials "WAC," indicating he had "worked all continents." QSL of B. T. Simpson, provided by the Southern Methodist University Amateur Radio Club.

tative of the sender" (see figures 2.2 and 2.3). One article offering design suggestions for QSLs instructed that the overall appearance should be "workmanlike" and warned against color combinations that "would lack strength" or "appear garish and cheap."[29]

To satisfy curiosity about "what kind of face goes with the voice or fist" heard over the radio, hobby publications often recommended putting photographs on confirmation cards.[30] Traditionally such postcard photos showed a ham seated alone at the operating position in his radio shack (see figure 2.4). The subject matter of photographs hobbyists sent separately varied from this pattern. Amid dozens of snapshots, mostly from the 1940s and 1950s, that one ham received following on-air exchanges, just a few included radio equipment and shacks. The vast majority depicted only human subjects—the hobbyist, and sometimes his wife and children.[31] Enclosing a family photo in a letter had the potential to broaden a budding friendship from its initial focus on radio and at the same time confirmed the sender's heterosexuality, clarifying the limit of this new relationship between men.

Figure 2.3
A hobbyist in Hawaii included a drawing of a hula dancer on his radio contact confirmation card, along with a logo designating his membership in the American Radio Relay League. QSL of Roy T. Kobayashi, provided by Thomas Roscoe, K8CX.

Figure 2.4
K3UOC sent confirmation postcards in 1964 that showed what he looked like while operating his amateur radio station. QSL of Mike Manafo, printed with his permission.

The space on confirmation postcards was largely reserved for technical data and limited hams' correspondence on QSLs. To make up for this, one handbook explained, many hobbyists sought "personalized and expanded communications." Another guide suggested that hobbyists include "letters describing their station in more detail and setting up schedules [for future conversations] with the other operator" when sending QSLs. "The desire to truly communicate with distant lands rather than merely logging countries and exchanging QSL cards" inspired some to send magazines and other small gifts to friends they knew only from talking by radio. This type of contact, according to one hobbyist, constituted "meaningful" communications and brought "additional pleasures" to ham radio.[32]

Meetings in person, which hams called "eyeball contacts," solidified friendships begun on the air and through correspondence. The Sandia Base Radio Club in Albuquerque, New Mexico, sponsored a "Friendship Award" that functioned much like an off-air analog of the ARRL's award for "rag chewers." To be eligible, a ham had to contact twenty-five local hobbyists and follow these on-air meetings with eyeball contacts, documented with the new friends' signatures. Handbooks encouraged visits between distant hams by pointing out that staying with a fellow hobbyist when traveling "cuts down on expenses, and the hospitality is always first rate."[33]

Since mid century, hundreds of radio clubs have existed simultaneously in the United States, formalizing in-person gatherings between hams who lived near each other, worked together, or shared particular radio interests. The Los Angeles area alone had more than thirty clubs active in the 1950s.[34] Clubs grounded hobbyist values in a visible social unit and provided vital mechanisms for enculturation. Hobby publications described clubs as offering the structure that individuals needed in order to feel connected to the ham community. Of the eight benefits of membership the Rochester Amateur Radio Association advertised in 1953, five focused on the pleasures of being part of a group. The club offered "Participation in club events open *only* to club members" and "Enjoyable monthly meetings." For $3 a year, the hobbyist was told he could expect "Fraternity with fellow hams from all walks of life" and a sense of "Belonging, knowing you're associated, being a part of things." Should anyone question his inclusion in this community, the club member could answer the challenge

Figure 2.5
Members of the Southern California DX Club and the Mid-Cities Amateur Radio Club posed by an abandoned oil derrick near Compton, California, atop which they had mounted an antenna to use during the 1949 Field Day contest. Photograph by William Martin. Print provided by the Southern California DX Club.

by presenting his "Billfold-size membership card."[35] Similar comforts of community could be found in looser affiliations, too. Specializing in a certain type of radio operation, according to one hobbyist, offered "a new sense of identity—a sense of belonging" by defining a smaller sphere of interaction.[36]

Ham clubs devoted time to social activities only superficially linked to technical matters. Participation in "meetings and social affairs" received a boost as hobbyists lost contact in the transmission silence imposed by the FCC during World War II. But clubs' commitment to social events far outlasted the wartime shutdown. The Northern California DX Club, whose members shared an interest in making long-distance (DX) contacts, explicitly promised to keep the amount of ham radio business conducted at meetings "to an absolute minimum" in the interest of promoting "good fellowship." Downplaying the technology central to the hobby was a strange but common practice. In Rochester, New York, the chairman of a club explained that it existed for "holding informal dinner 'get-togethers' for the purpose of chewing the fat about DX." "The aims of the organization are not ambitious," he wrote in the early 1950s, and were "more social than political." One meeting agenda set aside a time specifically for a "Gab-fest." As proof of the "unity and interest of the group" in hosting social functions, the chairman cited the fact that "At no meeting has the attendance dropped below 75% of the membership." Friendly gatherings remained the Rochester club's focus nearly thirty years later, with "Business not [to] be conducted at the meetings except in special cases required by the constitution."[37] Surely ham radio was a popular topic of discussion among club members, but meetings were not group tinkering sessions. Clubs gave hams a chance to enjoy each other's company away from the technical constraints of radio.

In the relaxed atmosphere of clubs, hams were gradually socialized into the hobby community. *CQ* magazine called clubs "the seat of true democracy in amateur radio" and charged each to "keep 'working on' its new Novice licensees and help to make *good hams* out of them." This process required "a lot more than [lessons in] technical and operating proficiency, and includes indoctrination into organized amateur activity [...] and in the traditions of our game." As part of their cultural instruction, hobbyists learned and practiced radio jargon in clubs. A handbook for new hobbyists described the typical meeting as "mostly informal—much 'rag chewing' goes on, coffee and doughnut breaks are common, and ham jabber fills the air, much of which will rub off on you." Once the "gibberish" of hams' language began to "form a pattern," a newcomer could become "an enthusiastic participant" in meetings and other hobby activities.[38]

Newsletters captured the casual, friendly interaction of clubs. Typically these were monthly publications produced inexpensively by a volunteer editor. They were intended as "extremely personal publications in contrast to the commercial jobs," according to one editor, and aimed to "deal directly and personally with each and every member of the club, in name as well as in activities." Because hams took pleasure in "reading about themselves and about the folks they know," the audience for club bulletins tolerated amateur publishing efforts. The ARRL reassured editors intimidated by literary responsibilities that it was all right to "know more about gamma than grammer [*sic*]" since newsletters were "just another means of communication among friends—like ham radio." Club publications deliberately retained a local flavor and plain language. After one bulletin made improvements, it promised readers, "This does NOT mean that we are going high-brow, far from it. We will try to give you the same 'home' type of publication with personal news of the hams you know, all the latest dope on Club doings, everything in fact that it has contained before."[39] Every page of club newsletters, in style and content, displayed the culture of ham radio.

To explain the basis of that culture—from the expectations for behavior to the preferred manner of speaking—radio hobbyists always pointed back to their chosen leisure technology. Certainly many ham values derived pragmatically from wireless apparatus. Audible transmissions depended on precise operations, and open exchanges required discretion. Admonitions in ham publications against faults such as messiness had a more tenuous technical connection, though might still be plausibly justified with claims that, for instance, electronics performed more reliably when constructed tidily.[40] But some characteristics of two-way radio operators came to be perceived as based in the technology only as a result of considerable effort expended by hobbyists. Detailed examination of the development, propagation, and defense of ham radio's masculinity reveals how one extraneous trait became a component of the identity of ham radio and provides insight into the construction of the hobby culture generally.

Ham Radio Made Masculine

The men who dominated ham radio—outnumbering women by 19 to 1—used a combination of visual, rhetorical, and social strategies to portray the hobby as deeply masculine. Stabilizing the gender of radio was a complex and, at times, subtle process. In part, it took place along the way as hobbyists negotiated their relationships to the electronics industry, the military, and their families. This subject therefore trails throughout the following chapters. I concentrate here on the actions and attitudes hams took primarily for the purpose of making the hobby manly.

Precisely because ham radio operators so effectively represented the hobby as manly, it is important to note the feminine associations with wireless communication. Women in the workplace frequently operated and built the technologies central to leisure radio. Between a third and a half of professional telegraphers were female.[41] During and following World War II, electronics manufacturers, praising women's small hands as capable of intricate work, commonly employed female assemblers. Women constructed most of the commercially available ham radio equipment sold by two leading producers, Hallicrafters and Collins Radio. The Collins Radio Company first hired women in its production department in 1942. Within two years, women gained a majority among its assemblers. The workforce at Collins became increasingly feminized after the war, and by the early 1970s women made up the vast majority of the assemblers at the main Collins manufacturing plant. In the mid 1950s when the Hallicrafters Company promoted "the six scientifically-designed production lines" at its recently modernized plant, all two dozen of the electronics assemblers pictured were female.[42] Men's leisure telegraphy and radio equipment construction risked looking like women's work. Additionally, passing time in casual conversations was stereotypically viewed as a feminine activity.

While fashioning ham radio as masculine, hobbyists also needed to clarify that this was a heterosexual masculinity. The pairing of men in a 1952 advertisement for a radio component illustrates how easily hams' fraternity could have been mistaken as homoerotic (figure 2.6). General Electric claimed that with its new power amplifier "a whisper will make your speaker roar." As representatives of the hobby radio community, it only made sense that both people depicting the whisper were men. Hams would have understood unambiguously that the men were stand-ins for electronics components, but the imagery might have misled outsiders. The uncharacteristic act of men whispering, the puckered lips of the whisperer, and the raised eyebrow of the listener could have been interpreted as sexually suggestive or politically subversive, a secret exchange between lovers or spies. The early Cold War perception of a link between homosexuality and Communism made potential confusion about hams' relationships a double threat.[43]

After World War II, male hams deliberately advanced a masculine identity for radio hobbyists and radio technology. Hobbyists in the first part of the twentieth century had appeared manly as they experimented with new technology and conquered the unexplored frontier of the airwaves.[44] But as two-way radio lost hi-tech status, hams needed to cultivate a masculine image for the hobby. A 1981 manual went as far as to promise readers that

6BK5

9-Pin Miniature Beam Power Amplifier

Heater voltage, a-c or d-c	6.3 v
Heater current	1.2 amp
Max plate voltage	250 v
Max plate dissipation	9 w

GANGWAY FOR HIGH SENSITIVITY plus real audio wallop! With just a few volts' drive, the new 6BK5 will put out enough power for BIG speaker response!

COMPARE the 6BK5's 3.5-w output from 5 v drive, with performance of these well-known audio types:

6AK6	requires 9 v drive to put out 1.1 w
6AS5	requires 8.5 v drive to put out 2.2 w
35L6GT	requires 8 v drive to put out 3 w
6BF5	requires 7.5 v drive to put out 1.9 w
35C5	requires 7.5 v drive to put out 1.5 w

YOU'LL FIND MANY WAYS to apply this new G-E marvel that hears more, and repeats it loudly! A 1-tube phono-amplifier for the junior op.—the power-output stage of your new receiver—an improved circuit for your home-entertainment amplifier—these are among the numerous possibilities.

YOUR G-E TUBE DISTRIBUTOR is ready with price and further facts. See him today! *Tube Department, General Electric Company, Schenectady 5, New York.*

RADIO MILESTONE:

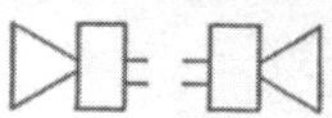

Tubes for Loud-Speaker Reception

● In 1925, General Electric developed the first special radio tubes for loud-speaker reception. This was but one of many pioneering steps by which wireless was freed from earphones, to take its place as today's foremost communications and entertainment medium. The long succession of General Electric radio-tube "firsts" is background for the superior design of advanced types like the 6BK5 which you buy and install in 1952!

ELECTRONIC TUBES OF ALL TYPES FOR THE RADIO AMATEUR

GENERAL ELECTRIC

166-188

the experience of completing a long-distance radio contact brought with it "the same power one feels with superbly engineered cars, strong horses or the ability to influence others." Generally, though, hobbyists based radio's masculinity in what they called "technical mastery" rather than attempt to tie radio to physical activity and strength. In the world of radio, a man proved himself with esoteric knowledge of electronics theory and with manual skill demonstrated by soldering components together to construct equipment. A 1957 hobby handbook actually referred to "the manly art of building your own receiver." Active engagement with technology exhibited a form of control and power suited to the age of electronics. When suggesting in 1968 that "a modern touch of excitement and challenge" might attract new hobbyists, the editor of *CQ* magazine acknowledged that this technical lure was unlikely to tempt rugged types like "the surfer or the skier." But he believed it had potential if pitched to a broad audience through the popular press and, he added, "yes, even through *Playboy*." A similar drive for "image enhancement" a decade later plainly stated the limitations on the hobby's style of manliness. "While it is unlikely that Amateur Radio will ever equal the Macho image of the hard-driving trucker with his CB mike in hand," wrote the editor of another hobby magazine, "we can certainly benefit by a public image that gets us out of the category of 'recluse tinkerers!'"[45] Ironically, outsiders' view of hobbyists as "recluse tinkerers" had resulted in part from hams' trumpeting of technical mastery.

Imagery in hobby publications reinforced the association of radio equipment construction with masculinity. The pattern in advertisements of the second half of the twentieth century was to depict technical ability as a manly trait but feature women models when displaying technology. The representation of disembodied, clearly gendered hands in RCA's free periodical *Ham Tips* demonstrates how one manufacturer adapted these conventions to suit ham culture. RCA displayed a new vacuum tube, which

Figure 2.6
Outside of the ham radio context, the image of one man whispering into another's ear, the puckered lips of the whisperer, and the raised eyebrow of the listener could have been misunderstood as sexually suggestive or as representing an act of political subversion. General Electric advertisement, *QST*, August 1952, page 1.

was not yet available to hobbyists, pinched delicately between a woman's perfectly manicured thumb and forefinger. For parts being sold for ham radio construction projects, however, a man's hand was used. The advertisement text explained the gender switch by describing these electronics components as "created by the hand of experience," suggesting that the same technically skilled, male hand displayed them. The rough, strong hand matched the parts' "rugged construction" and was large enough to grasp a selection of three tubes at once. The general illustration practices of *Ham Tips* maintained the image of male hobbyists as technically active. A photograph accompanying an article on "The Make-Your-Own Microphone," for instance, showed the microphone in the male hand of the builder.[46]

Hobbyists spoke from a position of technical authority when recounting instances of radio operation interfering with television and frequently named women ignorant of electrical matters as the source of interference complaints. Typical was a hobbyist's portrayal of hysterical housewives preemptively spreading rumors about electrical disruptions after a ham installed a new antenna: "Immediately Mrs. Jones down the block informs Mrs. Smith who lives next door that said 'contraption' has certainly interfered with their broadcast reception, even tho' the ten meter beam may not yet be in use." Another story of this type referred to an "old bat" who blamed hobbyists for imperfect television reception unrelated to electrical interference. One article's suggestion for stifling a female neighbor's nagging about interference was to deal directly with her husband. In a conversation between men, an angry neighbor might soon admit that the interference was really only on one channel—a channel he did not even watch and whose cooking show had given his wife despicable "recipes for braised moose jowls."[47] Radio hobbyists who dismissed female neighbors' accusations of interference in this manner contrasted technically knowledgeable men with women who frivolously watched television.

Hams implied that electronics know-how granted them an elite and manly status by characterizing the community that formed over the airwaves as a "technical fraternity." This fraternal social structure was made more apparent when the ham community organized into clubs. In some cases, ham radio clubs looked much like gentlemen's clubs. The Northern California DX Club (NCDXC), founded just after World War II, was all male until 1963 and accepted only three female members in the next decade. One member attributed the small number of women in the club to the fact that it focused on long-distance operations, which he felt was a disproportionately male specialty within the hobby, but acknowledged that club activities likely also contributed to keeping down the number of female members. He called the NCDXC a "smoker type club" whose gatherings resembled those of "a men's social club." Its meetings always included dinner and drinking and sometimes were held in venues that offered entertainment such as the "Mermaid act" members enjoyed at a meeting in 1961. In the early 1970s, some NCDXC members voiced concern about raucous behavior at meetings. The club newsletter, *The DXer*, published a complaint that "the members only want to get together for a drink and meal, and don't give a damn about DX, anymore!" In the next issue of *The DXer*, the editor pleaded with members to "keep the booxin [boozing] at a reasonable level" and "create a more friendly and cooperative atmosphere throughout the meeting."[48]

Much more so than the monthly outings, the annual joint meetings of the Northern and Southern California DX Clubs gained a reputation for debauchery. Reports in *The DXer* during the 1960s and 1970s repeatedly alluded to excessive alcohol consumption over the course of the several days the clubs met together in Fresno. The Northern club equipped the bus chartered for the 1963 gathering with a "portable bar, 50 pounds of ice cubes and 100 paper cups" for its 29 passengers. After the 1967 joint meeting, the editor of *The DXer* "suppose[d] that a great many of us have but an indistinct and foggy memory of what occurred in Fresno." The antics in Fresno drew attention throughout the ham community, including the report by *CQ* that "The traditional cocktail party made good use of four bars" at the hotel where the clubs met in 1973.[49]

The NCDXC newsletter hinted that sexual liaisons were one reason club members traveled to Fresno. The two host clubs extended invitations only to "DX guests." In particular, non-hobbyist spouses did not accompany hobbyists to Fresno, although it was routine for them to tag along to other ham radio conventions. The Northern club's newsletter promoted as a special point of interest for the joint meeting that the Southern club had more female members. A snapshot printed in *The DXer* of a man smoking a pipe and talking to a woman holding a drink was captioned, "Er well what're you doing after the dinner?" The man was not identified, but the woman's call sign was given. Other comments mentioned dalliances with women from outside the hobby. At the 1973 meeting, a naked woman jumped out of a cake as a way of rewarding one ham "for his long service to the Fresno meeting." The event was described in both regional and national hobby magazines, and a year later the *DXer* newsletter printed a photograph of "Brandy" hugging the meeting organizer as a way "to remind" members of the next joint meeting. *The DXer* further suggested that when the club traveled to Fresno some members took advantage of a weekend of anonymity to pick up women and visit prostitutes. One cartoon on the cover of the 1974 "Fresno Issue" of the newsletter showed a woman in a strapless dress telling a man at a bar, "It's getting late and I'd like to go to bed before I get tired." In a second cartoon, labeled "Awards," the plaque a ham brought back from Fresno confused his wife. "It's for perfect attendance at Madame Zelda's what house?" she quizzed him.[50] Though it is difficult to judge the extent to which these remarks accurately represented experiences, there must have been at least a grain of truth to the portrait of the Fresno meeting as a freewheeling men's retreat or the humor would have fallen flat.

Alongside ham clubs that fit the men's social club formula were radio clubs modeled upon fraternal community service clubs. After World War I, veterans' groups and general service clubs became more popular with men than strictly social clubs.[51] In line with this trend, mid century radio clubs commonly emphasized their similarity to organizations such as the Jaycees, Rotary, and Lions Clubs and publicized their involvement in civil defense drills and in providing communications following natural disasters. Ham publications encouraging participation in such service activities openly stated that one motivation was to improve the hobby's public image, "to protect and promote ham interests on the local level."[52]

Structuring ham fellowship as a fraternity facilitated alliances between radio clubs and male service organizations. Men who were both hams and Rotarians, for instance, described an ideological affinity between the communities including "the easygoing informality typified by the use of first names, the willingness to serve others, the development of friendship across international borders and among men of varied occupations." As evidence of shared attitudes toward civic responsibility, one member of both groups noted that the Rotary's All-American Award for Public Service twice had been given to ham radio operators. Rotarians made use of ham radio to "broadcast messages of brotherhood to every continent" in an act of "Rotary fellowship." "Surely if 'there is a destiny that makes us brothers,'" wrote a Rotarian ham, "amateur radio is doing its bit to realize that end." When a Rotary club in Hawaii met over the airwaves with one in New Mexico, the organizer of the ham radio linkup called it "more than just a stunt." The meeting "joined two Clubs in closer fellowship," he claimed, and offered "a better understanding between two distant parts of the world, albeit parts of the same country," leading him to conclude that ham radio "was definitely 'good Rotary.'"[53]

The camaraderie of radio hobbyists with fraternal groups spanned the generation gap. In 1949, the trade journal *Radio and Television News* pronounced the association between hams and the Boy Scouts to be "to their mutual advantage" and gave the example of a Chicago scoutmaster offering a course on ham radio to distinguish his troop. The "educational appeal and glamour" of the hobby drew forty-five boys and nine fathers to the first class. The Boy Scout organization hoped radio would teach boys electrical skills as well as elements of hams' technical culture. Advice on earning a merit badge in radio stressed the importance of deep understanding over the ability simply to follow directions. Unlike assembly-line workers who could say only "that they soldered a 'couple of wires to a thing,'" scouts were to learn "the reason for each connection, each part." The Boy Scouting suggestion to "poke your nose around until you find out *why* one wire connects to the detector tube and another to the capacitor" echoed ham radio handbooks' encouragement of tinkering and learning by doing. Like the Rotarians, Boy Scout leaders believed ham radio could enhance fellowship. The handbook for the radio merit badge mentioned the possibility that, because many hams had found their way into the hobby through scouting, radio contacts would provide introductions to other current and former scouts.[54] The Boy Scouts adapted particular components of the ham technical fraternity to scouting. Their magazine *Boy's Life* sponsored a radio club and held "Hamborees" that combined the style of the Boy Scouts' Jamboree gatherings with that of radio hobbyists' "Hamventions." Merit badge requirements directly linked Boy Scouts and hobbyists by allowing only licensed hams to judge a scout's knowledge of Morse code and electronics theory and to inspect the basic receiver the boy had constructed. Radio clubs encouraged members to befriend scouts and supervise examinations. From the perspective of radio hobbyists, the Boy Scouts—a fraternal, militaristic organization focused on character building and community service—seemed like a perfect membership feeder group. The ham community suggested that radio enthusiasts recruited from local high schools, religious youth groups, and especially the Boy Scouts could provide "the transfusion that ham radio needs."[55]

Whether joined officially into fraternal clubs or just loosely banded together as a technical fraternity on the airwaves, the hobby radio community was distinctly masculine. Male hams used a playfully harsh rhetoric to prevent feminine infiltration. *CQ* magazine's short report about the 1971 national Hamvention exhibited three common strategies for policing the gender border. First, females who showed an interest in ham radio were portrayed as a threat to the hobby. A photograph of a young girl at the convention—the picture of innocence, wearing flowered shorts and her hair tied into a ponytail with a ribbon—carried the caption: "We're not sure whose liberation front sponsored this one, but they wouldn't let her join." Second, hobbyists depicted wives as spoiling men's fun by interrupting technical leisure with family obligations. A television news story about the Hamvention had picked up on this sentiment and reported that the pastime drove a wedge between spouses. As an ironic counterexample, the magazine ran a photograph of a couple at the convention with a caption stating that the hobbyist "would rather not talk about" the fact that he had brought his wife along. Finally, hams frequently described feminine sexuality as competing with technical apparatus for men's attention. The convention report attributed the popularity of a slow scan television demonstration in part to "the Hot Pants-ed, redheaded circulation femme" who was distributing leaflets for a neighboring booth.[56] Joking tempered, but did not obscure, the message that women interfered with ham radio culture.

The small group of women who did become radio hobbyists faced a dilemma. To be accepted as true hams, women had to take on a technical identity. The close association male hams had forged between radio technology and masculinity, however, meant that in attempting to enter the ham community women risked undermining their femininity. It is an indication of how strongly gendered ham radio was that female hobbyists in this difficult position responded by emphasizing stereotypically feminine traits and behaviors while downplaying technical interests and abilities.

Amelia Black, author of the first women's column to appear in one of the ham magazines, projected a hyper-feminine image of herself and the women she wrote about. Black noted, for instance, her disappointment at winning a piece of radio equipment instead of the pair of nylons offered as a door prize at a meeting of women hams. And she called her membership in the Rag Chewers Club—the group that encouraged lengthy, non-technical conversations on the air—something that "might be expected" from a woman ham. Eleanor Wilson, who held the equivalent of Black's position at a rival magazine, described women as keeping their distance from ham radio competitions. Normally contests focused on manly demonstrations of speed, technical power, or domination of the airwaves by challenging participants to make the most contacts, reach the farthest distance, or talk to hams in the greatest number of countries. In those cases, Wilson always felt it necessary to deliver a "sermon on contest spirit." But with the approach of a contest that was more a social mixer than a competition, in that it only counted radio exchanges made between a man and a woman, Wilson teased her readers, "Strangely enough, you never need much coaxing for this party."[57]

Descriptions of women hams in the popular press carefully pointed out the feminine touch they brought to the hobby. A 1942 newspaper story mentioned that one woman's housekeeping skills kept her hobby space free of "the mess usually associated with amateur radio." *Recreation* magazine characterized another woman's on-air activities as mostly "chatty conversations" and facetiously declared it an act of ham radio heroism that she had once supplied a ship stuck in the ice near Greenland with a cornbread recipe. When *Time* named seventeen notable hams in 1961, the list included just one woman. The magazine's only statement about this recent Miss Universe was "35-23-35," dressmaker shorthand for her bust-waist-hips measurements.[58]

The women who participated in ham radio clearly had technical interests and skills, or they never could have passed the licensing examination. Yet female hobbyists often dismissed any technical identity, apparently in an attempt to protect femininity. Amelia Black's debut column in 1946 promised that technical advice would appear only if readers insisted. The first woman profiled there brushed off her own technical knowledge as "slight," saying she considered herself "fortunate that there's always been a brother or a husband to build the equipment and to keep it perking." In the pages of ham magazines, women hobbyists continued to deny their technical engagement and to emphasize feminine practices through the 1970s. In 1976 one female hobbyist wrote, "Oh, there are a few women here and there who are hams, but they don't seem to do anything. They just talk." She blamed her own failed attempt to assemble a piece of equipment on child care responsibilities, which she felt were "part of the reason why women generally accomplish so little in amateur radio."[59]

Portrayals such as these served the double function of neutralizing the power of women's technical ability and shielding women hams from the accusation that their technical hobby made them mannish. The strategy of stressing femininity in order to make women's skills appear less threatening was not confined to ham radio nor to technical fields. During World War II, the press described women in the computational jobs previously held by men as engaged in "'domestic' work for the nation." After the war, women engineers were depicted as stereotypically feminine to mark a distinction from the physically strong image of Rosie the Riveter. Popular magazine articles about successful business women in the first postwar decade used related tactics to cordon them off from the business world dominated by male achievers.[60]

Preserving their gender identity prevented female hams from attaining genuine insider status in the hobby community. Ellen Marks reported experiences in the late 1970s as a woman "in a 'masculine' hobby" that were little changed from those of women in the 1940s. Men who saw Marks at ham radio conventions assumed she was just there with someone and not a hobbyist herself. Sending Morse code on the air hid her appearance and voice, so fellow hams commonly assumed Ellen Marks was a man and replied to her as "Allen," thinking they had de-coded her name wrong. Once Marks corrected these gender misinterpretations, male hams tended either to ask her about cooking or suppose that an interest in "sports and mechanics" must be paired with her ham radio hobby.[61] Women hobbyists simply did not make sense amid a community that so valued masculinity and brotherhood.

Radio clubs in particular were reluctant to take on female members. Many service oriented clubs like the Schenectady Amateur Radio Association shared the demographics of gentlemen's type clubs like the NCDXC: nearly thirty years after they were established, each of these clubs had more than a hundred members and included only three women. A 1945 hobby magazine article about the North Shore Radio Club of Long Island claimed that women were "by no means excluded," but the accompanying photograph showed a club event perfectly divided along gender lines. With a caption describing the women members as "properly segregated and tabled," the overall message seemed to be that clubs welcomed women as long as they kept to themselves.[62] Confronted with men's control of most clubs, female hams formed clubs of their own.

Within mixed gender clubs, women encountered stereotyped views of their proper roles and skill. This cannot be blamed entirely on men. In a 1960 article with the alarmingly misogynistic title "'Club' Your Women," Carole Hoover detailed the many submissive roles women might play in radio clubs. If women hams followed her behavioral guidelines, Hoover believed, a feminine presence could improve clubs while continuing to "let the fellows run the show." Hoover explained that she and a handful of other women, "Going on the old adage that the 'way to a man's heart is through his stomach,'" had gained entry into the local club by taking refreshments. What she called the "girls' meet-and-eat plan" led to the women catering weekend contests and holding bake sales to support the club. Along with doing the cooking, Hoover suggested women handle club paperwork, declaring the offices of secretary and treasurer "jobs that women just simply enjoy more than men [do]." But, Hoover warned, women needed "to guard against a 'take over' impulse that could shrink a thriving club to a sewing circle in a matter of weeks." Clubs were to remain masculine. Hoover told women to forget the idea of prettying up the club meeting place, although she found it acceptable "to empty wastebaskets, sweep up, and sneak off with a 'girlie' calendar now and then."[63]

The positions held by the first women to join the Northern California DX Club demonstrate that female hams employed at least one of Hoover's techniques to make a place for themselves in clubs. When the NCDXC accepted its first female member in 1963, a handful of the men present stood up and walked out of the meeting, never to return. But just three months later the new inductee proved useful to the club by taking on the job of secretary/treasurer. A decade later, the third woman to join the NCDXC also volunteered to serve as secretary within her first year of membership.[64] Taking on administrative responsibilities earned women official inclusion in clubs but did little to integrate them into the masculine culture of ham radio.

Through the combined strength of countless tiny gestures, hobbyists marked two-way radio as unquestionably manly. This had a number of social benefits for hams such as reinforcing their gender identity and generating a sense of fraternity within the ham community and with other men's organizations. Radio hobbyists of mid century were hardly alone in endeavoring to stabilize perceptions of technical activities as suitably masculine. In a Cold War context that conflated techno-scientific know-how with international political power, elite hi-tech organizations also paid attention to technology's image. NASA, for instance, embarked on a public relations campaign to depict the astronauts on early missions as active commanders rather than just submissive passengers. Meanwhile Stanley Kubrick's film *Dr. Strangelove* caricatured combatants in scientific warfare as disengaged button-pushers.[65] The perceived identity of a technology affects attitudes toward technology, which in turn affects critical decisions about investment, regulation, consumption, and implementation.

The centrality of technology in the ham community inspired a cycle of social-technical identifications. In one step, radio hobbyists granted technology a social identity by establishing guidelines for the use of two-way radios as well as for off-air interaction within a group defined by its shared interest in two-way radio. As hams articulated, taught, and enforced expectations for behavior with regard to radio, they associated social norms with the technology. Hams routinely implied, and occasionally explicitly stated, that radio required operators who possessed traits such as precision, efficiency, discretion, rationality, attentiveness, political neutrality, and masculinity. Their logic was that these traits were technical demands of radio, following from the way devices were constructed and how they functioned. In the other step of the cycle, hobbyists took on a technical identity from radios. By personally identifying with technology and as radio operators, hams reflected back onto themselves the very characteristics they had imparted to radio technology. This cycle produced the culture of ham radio.

3 Equipping Productive Consumers

Hams were always active. Talking on the air—ostensibly the primary hobby activity—was the most sedentary thing hams did. When they were not rigging up antennas or hauling gear to the top of a hill with the hope of making distant contacts, they were building, testing, repairing, and modifying equipment at the workbenches in their shacks. The ham community took special pride in construction feats that displayed an extraordinary level of technical mastery. Heroic stories told of hobbyists with a radical do-it-yourself spirit who blew the glass to make their own vacuum tubes or who ground their own tuning crystals. But hams did not set independence as a standard or even an ideal, and they routinely purchased supplies and equipment. In 1938, Theodor Adorno ridiculed the radio amateur as a pawn of the marketplace because "He patiently builds sets whose most important parts he must buy ready-made."[1] Adorno's criticism misrepresented the character of radio commerce as greatly as did tales of homemade vacuum tubes. For while the hobby fundamentally depended on the radio-electronics industry, consumption did not prevent hams from being productive and creative.

Like all hobbyists, amateur radio operators existed in an intermediate position between producers and consumers as they pursued activities that straddled the commonly juxtaposed categories of labor and leisure. Hobbyists bought tools, parts, kits, and scaled-down or somewhat simplified versions of professional equipment. They then used these goods to produce their own creations.[2] In technical hobbies, this pattern of productive consumption altered the style of technology that manufacturers sold to customers. Most twentieth century consumer technologies, designed to make hi-tech machines accessible to the largest possible audience, concealed functional mechanisms and reduced user requirements from skilled engagement to superficial button-pushing. Quite different devices emerged from the unusual producer-consumer relationship between ham radio manufacturers and hobbyists.

To Build or To Buy

The ham radio community saw building equipment as a perfect opportunity for learning by doing, the form of education that amateurs favored. With no trace of irony, the hobby literature tried to instill in readers the value of hands-on learning over book learning. Handbooks across the decades claimed that "The 'know-how' obtained by constructing a piece of electronic gear cannot be duplicated by reading a thousand books!" (1957), and "The knowledge of electronic fundamentals obtained from doing cannot be duplicated by reading textbooks" (1982). To allow hobbyists to learn from experience, manuals guided novices through scripted building projects that taught basic skills and the habit of tinkering. Equipment construction became so closely identified with ham radio culture that a 1973 hobby electronics handbook called hams "the true hobbyists who started the build-it-yourself concept in electronics."[3]

Construction by radio hobbyists dropped off as its financial advantage shrank and the selection of ready-made equipment grew. In the first half of the twentieth century, few amateur radio models were available for purchase, and those that were came with hefty price tags. For example, the Hallicrafters HT-1 transmitter sold for $195 when it debuted in 1937, at a time when a complete amateur station could be built from parts for a quarter of that cost. By the late 1950s, the cost savings of building equipment had diminished and hams could purchase a wide variety of specialty gear. One handbook then described hobbyist-built receivers as "almost as extinct as the famous dodo bird," a reversal of the situation "A decade or so ago [when] the home-made receiver was the rule rather than the exception."[4]

Whenever building appeared to decline, the ham community expressed anxiety about the survival of the hobby. A 1959 manual called the "trend" for buying ready-made gear "detrimental to both the ham and the whole field of ham radio." The author explained how this presented a threat to the hobby culture with the accusation that "new operators who do not build their equipment violate the spirit of technical curiosity and achievement which has been the foundation of amateur radio."[5] Appeals to community responsibility were common in attempts to encourage building. The importance of home construction was a topic that *CQ* magazine "harped on again and again" in the 1950s and 1960s. Announcing a "$1000 Cash Prize 'Home Brew' Contest" in 1950, *CQ* called independently built equipment "the type of gear which has helped to make amateur radio our greatest reservoir of technical proficiency." The magazine tried to steer hams back into building by sponsoring such competitions and by publishing more construction plans. Lessons traditionally learned by doing imparted a powerful technical mastery to hams. "As our ranks of home constructors thin we also fall to a lower technical level as a group," the editor worried in 1958.[6] Amateurs feared that if ham radio lost status as a technical activity, they might also lose the privilege of operating on the public airwaves.

A desire to be associated with technology was part of what lured hams to purchase equipment. Hams wanted machines that were outwardly hi-tech. Conceding that the "shortcoming of many homebrewed projects is the lack of finishing touches which add a commercial touch of glamour," construction handbooks recommended that hobbyists spend "an extra hour for the final cosmetics." Specific suggestions for achieving a "professional metal appearance" included covering wooden cabinets with metallic paint and purchased decals. But these crude approximations failed to replicate the sleek exteriors of commercial electronics equipment. In addition to delivering better looking machines, manufacturers produced extremely compact equipment by using smaller parts and placing them in tighter configurations than were convenient in home building. A 1971 review of an amplifier built with integrated circuits justified its high price by stating that amateur builders could not match the ready-made device's minimal size and weight.[7] Better performance rarely was mentioned as a reason to buy rather than build equipment, perhaps to avoid an affront to hams' pride in their own construction abilities.

The tension between the desires to build and to buy equipment complicated the amateur radio market. On the one hand, hams respected do-it-yourself principles and wanted to display technical mastery and to reap the rewards of learning by doing. These factors argued for constructing equipment in home workshops from original or published plans. The attraction of machines that looked and performed like the latest hi-tech, professional gadgets tempted hobbyists to consider purchasing ready-made equipment. To court ham radio customers, manufacturers mediated between these conflicting wishes.

The Market for Amateur Radios

The motivation for radio-electronics firms to participate in the small, specialty trade in hobby radio supplies is not immediately obvious. Bigger profits were found selling broadcast radio receivers, non-hobby two-way communications equipment, televisions, and computing technology. The relative strength of the markets for these other products indeed figured in manufacturers' decision of whether to sell to hams. In many ways, though, ever since the time radio was new and hobbyists were essential customers, the demand for other electronic communications technologies actually increased the appeal of the ham radio business.

There were two kinds of radio hobbyists in the first third of the twentieth century. Some sent and received signals; others listened to amateur and commercial transmissions without responding. Initially the radio parts and receivers used by the "transmitting amateurs" were nearly identical to those used by "broadcast listeners." The two "types of amateurs" represented a unified, thriving sector of the developing radio industry.[8]

All radio listening through the 1920s demanded technical skills for tuning, tinkering with, and constructing equipment. Given our familiarity with single dial or push-button radios, especially those which employ automatic frequency control or digital tuning, we easily can overlook the talent once required just for precise tuning. The clear reception of a station on an early radio testified to the abilities of both radio operator and equipment. Enthusiastic listeners kept log books recording which broadcasts they had received, when, and under what conditions. Into the late 1930s, radio stations sent "verified reception stamps" or postcards to listeners who wrote in saying they had heard a particular broadcast. Radio listening hobbyists preserved their ethereal radio accomplishments by pasting stations' stamps and cards into scrapbooks, including albums produced for this purpose.[9]

The expense of ready-made receivers inspired some consumers to attempt building their own. Like many other handbooks available to guide novices, *Radio Construction for the Amateur* (1924) offered thorough instructions for building a receiver from commercially available parts, including lessons in the basic skills of soldering and drilling. The cover illustration of a man wearing a crisp white shirt and smart vest while working on a radio showed that even those who had the means to purchase a receiver might choose home assembly, perhaps to have more control over the design. *Radio Construction for the Amateur* described numerous receiver models in detail and provided schematic diagrams for "eighteen popular hookups."[10]

Listeners who bought radio receivers still faced some assembly before they could spend evenings searching the dial for broadcasting stations. The instruction manual packaged with the Paragon Receiver in 1923 indicated that to bring sound through the headset the user needed to start by wiring in a battery and antenna. The Paragon Receiver included neither. Instead the manual explained how to build an antenna out of parts available from a radio supplier. A list of likely solutions guided the owner through common problems with wiring, and the manual provided advice on tuning techniques that would be useful once the radio was finally complete and ready for operation.[11]

As eager manufacturers began to compete for a stake in the booming radio industry, they accommodated the majority of the public, who preferred listening to tinkering, by dropping technical expectations. The radio industry had total sales of only a few hundred thousand dollars per year at the start of the 1920s. By decade's end, this figure had increased to nearly $750 million annually.[12] Mass production increased the supply of receivers and lowered the price, while critical shifts in the technical culture of broadcast listening bolstered demand. To attract a wider and particularly a female audience, mass-produced receivers of the late 1920s featured aesthetically pleasing cabinets appropriate for polite living rooms and an interface that required little more than tuning. Ready-made radios were so simple to operate that manufacturers no longer taught consumers how to work on the devices and began to discourage tinkering with the threat that it would void the warranty. The radio industry then professed an obligation "to protect the public" from electrical dangers.[13]

With the simplification of broadcast receivers, producers effectively split the general category of amateur radio into two forms of leisure, broadcast listening and two-way communication. Broadcast listening changed from an active hobby into passive recreation. Two-way radio communication alone retained the status of a hobby and the name "amateur radio." Hams did not necessarily dislike broadcast radio, and many were avid listeners. Yet ham radio always had been distinct from listening, and this distinction became more profound after the 1930s. Then tinkering with radios—the challenges, busyness, and productivity that made radio a technical hobby—thrived only in two-way radio.

The "transmitting amateurs" had begun the radio craze and formed the basis for the radio industry. In addition to starting the demand for parts and equipment, hams developed the technical culture of radio in a way that contributed to the creation of commercial broadcasting. Economist Hiram Jome acknowledged the importance of hams to the radio market when he used the number of licensed amateurs to estimate the demand for broadcast receivers in 1925. The logic of Jome's analysis hinged on hams being indirect lead consumers—not simply initial purchasers who led others to do the same, but consumers of one group of products who inspired others to purchase a related group of products. With their leisure transmissions, hams gave non-transmitting hobbyists something to listen to, suggesting an identity for radio as a form of entertainment that gave rise to commercial radio stations. RCA president David Sarnoff in 1930 noted that "the radio industry outgrew its first customer—the radio amateur."[14]

The direct contribution of ham radio parts and equipment sales to the overall radio business was minor for all except the very first years of radio's existence. Already in 1926 a radio handbook author figured that broadcasting listeners "easily outnumbered by one hundred to one" the transmitting amateurs. Radio manufacturers generally shifted their focus away from ham radio as radio listening rapidly grew in popularity. When the newly founded Chicago Radio Laboratory advertised its first ham radio receiver in 1919, it expressed a commitment to "what the amateur wants and needs." But in less than a decade the company transferred all its energies to selling broadcast and shortwave receivers and dropped the name "Chicago Radio Laboratory," which evoked tinkering and experimentation, for the more broad appeal of "Zenith."[15]

The lure of ham radio manufacturers into related, more profitable areas eventually even swept up two of the most beloved producers, Collins Radio and Hallicrafters. Arthur Collins turned his hobby into a small business after his father's industrialized farming company sunk into debt during the Great Depression. Two years after introducing the first fully assembled ham transmitters in 1931, financial success let the Collins Radio Company leave the founder's basement behind for more conventional factory space. Soon government and commercial radio buyers, frustrated by the limited availability of transmitters, started purchasing the equipment intended for amateurs. Meeting the needs of these unexpected customers drew Collins Radio into avionics and general communications. Producing ham equipment then became just a sideline, kept alive through the early 1980s partly as a matter of corporate tradition. The story at Hallicrafters was similar. Former Navy radioman William Halligan started Hallicrafters in 1933 to manufacture ham equipment. Though it quickly shifted production to concentrate on general-use shortwave receivers, Hallicrafters sold radios to hams until 1974.[16]

The smooth transition of these ham radio businesses to more diverse production indicates one way the expanding radio-electronics market of the 1930s and 1940s supported the hobby market. The technical foundations that two-way amateur radio shared with emerging electronic communications technologies made it relatively easy to manufacture the devices alongside one another. For this reason, dealing in ham radios was not merely a launching point for small companies. Even established radio industry giants RCA and General Electric pursued a sideline in products for hobbyists. Following years of selling vacuum tubes and other components to hams, RCA sold hobby transmitters and receivers for the first time in 1935. The annual reports for the next decade included amateur equipment among the list of "Principal Products of RCA," if often quite low on this list. GE supplied components to hobbyists, a decision validated at a December 1940 meeting of General Electric's "radio specialists."[17] The Radio and Television Department intended the proceedings, limited to a printing of 125 copies, to be confidential. A copy archived among the personal papers of one of the attendants provides a rare chance to witness a corporate evaluation of the hobby market.

E. E. Williams noted that GE sold to hams via "an entirely separate channel, namely, radio distributors." Any profit lost through this "somewhat unusual" commercial practice was believed to be sufficiently offset because distributors commanded "by far the largest percentage of the amateur radio business" and GE could reach distributors "without undue sales expense." The hobby trade had the further benefit of being "a replacement business which continues rather evenly throughout the year." "As a whole," Williams concluded, "the business from the radio amateur field is quite satisfactory even though of small volume."[18]

Figures presented by E. H. Fritschel supported Williams's general analysis. GE sold $2,812,000 worth of radio transmitting tubes in 1938. Hobbyists' purchases accounted for $400,000 of this total. Hams generated more revenue for GE than did four other sectors tracked by the Market Research Division, including police radio communications ($164,000). But greater sales resulted from the broadcasting ($862,000), export ($600,000), and government markets ($411,000). Hobbyist demand did not always rank so closely with government demand, which varied according to strategic operations, and Fritschel correctly predicted that sales to the armed forces in 1938 was "probably a record low."[19]

The modest, consistent business radio manufacturers had done with hams up until 1940 was resoundingly trumped by huge wartime demand. In the context of World War II—often called the "radio war"—two-way radio existed strictly for strategic purposes. The Federal Communications Commission (FCC) banned recreational use of the airwaves while the military specially recruited amateur radio operators and asked hams to donate equipment from their home stations for battlefield use. Whether ham radio or military radio, the underlying technology for two-way communications was the same.

Once again technical affinity facilitated flexible production. As manufacturers of all types of radios altered output according to the military's needs, the switchover proved easiest for the handful of companies that had sold to hobbyists in the 1930s. At Collins Radio, the most significant change was in capacity. Government funds for emergency manufacturing facilities helped Collins double the size of its factory in 1941. Hallicrafters had to make only minor modifications to existing products in order to complete $150 million worth of war contracts. Manufacturers of broadcast receivers undertook more difficult factory conversions to achieve efficient wartime production, followed by costly reconversions when government contracts disappeared at war's end. In the fall of 1945, *Business Week* predicted a bright future for prewar hobby suppliers, which "face[d] no reconversion problem."[20] The similarities between military and amateur radios thus brought about a paradox whereby ham radio equipment producers actually gained strength during the wartime pause of the hobby.

When the hobby of ham radio grew dramatically after the war, the demand for amateur radios took up some of the slack manufacturers experienced from reduced military demand. Wartime technical education acquired through military service or defense-related production enabled many more individuals to participate in electronics tinkering and other technical hobbies. Attention to a wider world—sparked by overseas tours of duty or years of listening to reports of distant conflicts—and the estimated two million soldiers and civilians who had become radio technicians during World War II contributed to the increased interest in ham radio specifically. Enthusiastic newcomers joined the prewar hams eager to return to the airwaves. As soon as the FCC resumed granting amateur radio licenses, applications streamed in at a rate that created a yearlong licensing backlog. The number of hams continued to rise for years to come, doubling over the next decade (see figure 3.1). New hobbyists needed equipment, the long-time hams who had donated radios for wartime use sought replacements, and others had pent-up purchasing demands, especially for gear that incorporated recent technical innovations.[21]

The government saw in the strong market for ham radios a chance to clean house and recoup a tiny portion of its wartime spending. During 1946, the military sold so much surplus radio equipment to the public that specialized resellers sprang up, and competition among these dealers drove down prices. Hams found the military's castoff rigs reliable and affordable. The machines also offered a connection to the war, nostalgic for those hobbyists who had served and novel for those who had not. Surplus radios demanded just the sort of technical engagement hams enjoyed. Civilian users had to restrict the power output of transmitters to operate within the legal limits for ham transmissions and had to modify military radios for

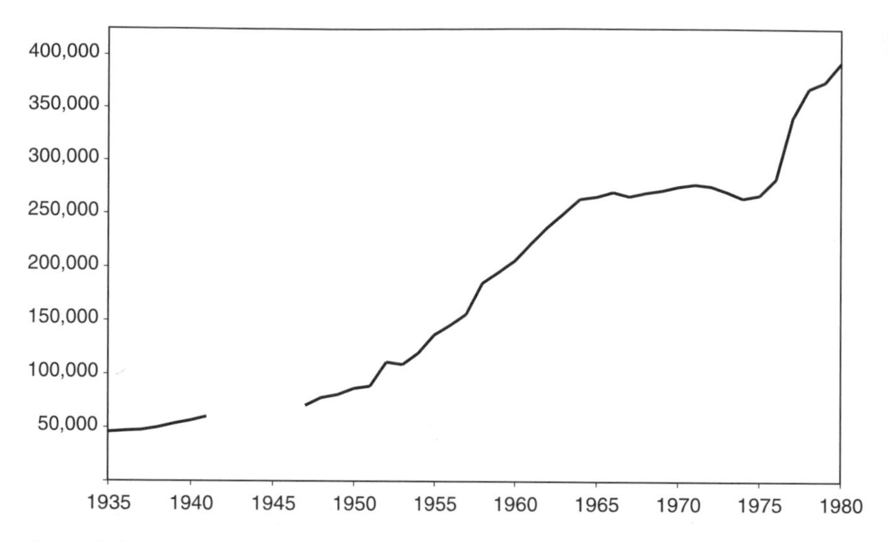

Figure 3.1
Number of amateur radio operators licensed annually by the FCC. Data compiled from U.S. Federal Communications Commission, *Annual Report* (1935–1980).

use on the frequencies set aside for amateurs. Many buyers made further adaptations for convenience, such as switching the main power source from battery to standard alternating current. Radio magazines and handbooks and the catalogs of surplus dealers provided schematic diagrams of military equipment, instructions for modifications, and general suggestions about the use of surplus gear in home stations. The hobby magazine *CQ* alone published 99 articles on surplus equipment in the first postwar decade.[22]

The demand for components grew along with the demand for equipment in the postwar period. All hams needed parts for modification projects and repairs, and many hobbyists still built equipment from scratch. From 1953 to 1956, sales of electronics parts to amateurs rose in value by 58%, and by 1959 nearly 700 parts dealers across the country sold ham radio supplies. Most of these distributors served electricians, engineers, and small industrial firms in addition to hobbyists. Stores specializing in amateur radio goods could survive only in regions densely populated with hams, such as Schenectady, New York, home of the major technical employer General Electric. Based on advertisements in the *Schenectady Amateur*

Radio Association Newsletter, at least four stores in Schenectady, and two more within a twenty mile radius, sold ham radio components and equipment during the 1950s. Two of the six were general electronics dealers. Another was a hardware store, which recommended that customers who had questions best answered by a fellow hobbyist "ask for 'Dale,' W2GRI." Electronics manufacturers and the infrastructure of supply houses mutually supported one another in carrying out sales to hobbyists. When GE committed soon after the war to "going into a vigorous campaign for the 'ham' business," it sent a memo to distributors to enlist their cooperation.[23]

The big story in the thriving postwar consumer electronics market, of course, was television. Despite the "accelerated" demand for amateur radio equipment, Hallicrafters's 1950 annual report showed that sales of television receivers accounted for more than five times as much of its income as did hobby radio sales. Still, sales to hobbyists generated a respectable revenue within the electronics industry, roughly $25 million annually in the late 1950s.[24] This represented a comfortable niche market, away from the higher-stakes competition for television customers.

Company traditions and personal sympathies seem also to have played a role in keeping electronics manufacturers interested in a small consumer group. A corporate history of Collins Radio repeatedly mentioned the nostalgia of managers for early product lines and for youthful hobbies as a reason the company continued to produce ham radios. Hobby equipment was such a "sentimental favorite" that Collins employees held a ceremony to celebrate the first ham gear released after the war. More than 40% of hams worked in the electronics industry, and it is possible that loyalties to their leisure pursuits influenced decisions they made in the workplace. Consider the example of E. E. Williams, who advised GE in 1940 to continue producing ham supplies. He was active in the hobby himself and four years earlier had compiled a directory of fellow radio amateurs/employees in an effort to strengthen ties between hams working at the Schenectady plant. Regardless of whether that network of hobbyists ever explicitly took a unified position on business matters at GE, the aggregate effect of each individual favoring ham radio would have been considerable. If promotional materials can be believed, hobbyists employed by electronics firms served as a ham's "personal representatives within the company" who would provide "Ham-to-Ham treatment."[25]

A combination of sound business strategy and a sentimental attachment to ham radio technology kept electronics manufacturers and parts distributors in the hobby market from the 1940s to 1960s. Then two interrelated changes weakened sales of ham radios. From the late 1960s to the early 1970s, the number of licensed amateur operators stagnated. The FCC recorded 266,000 hams in 1965, and only 1,400 more a decade later. (For comparison, this 0.5% growth over the decade followed two decades during each of which the number of hobbyists had doubled.) Only a dozen major amateur equipment manufacturers remained in 1967. Three years later, when declining advertising revenue forced *CQ* magazine to cut costs by using fewer color images and a lower quality binding, the publisher explained that "for some time now [. . .] from a business point-of-view the Amateur Radio industry has been in pretty sad shape."[26] The coincident transition to integrated circuits as the fundamental components of the electronics industry made ready-made equipment and kits less attractive to hobbyists, altered the hobby culture, and contributed to diminished enthusiasm for the hobby. It was a moment that clarified just how ham radios differed from other consumer electronics.

Technical Products for Active Consumers

Hobbyists needed productive, interactive electronics, not the style of mass-marketed consumer electronics that hid reliable technology inside visually pleasing cases with easy-to-use controls. With regard to technical specifications, manufacturers could have streamlined ham equipment much as they had broadcast receivers. But to deny hams a closeness to technology, and the deep theoretical understanding and practical skills that followed, would have demoted ham radio from an active hobby to a passive form of recreation on par with broadcast listening or television watching. The electronics industry instead tailored product design, sales tactics, and customer support to suit technically engaged consumers.

The understanding that hams would open machines pushed physical design considerations into the guts of hobby equipment. Advertisements for and reviews of ham radios in the 1940s and 1950s repeatedly highlighted two features of little interest to the typical electronics consumer, the appearance and accessibility of internal components. In promotional photographs of ham equipment, manufacturers revealed interiors to show the quality of assembly and that there was space to tinker easily. Nearly half of the photographs and illustrations of equipment in Hallicrafters's 1945 ham radio catalog exposed the insides of machines.[27] Providing this view was a gesture of manufacturers' respect for customers' technical knowledge, implying that hams could recognize good electronics.

Hams strongly associated orderly, precise appearance with sound construction technique. Often their judgment of these characteristics overlapped to create a kind of technical aesthetic. The caption to a photograph in a 1946 transmitter review, for instance, pointed out that the design "achieves unusual cleanness of appearance by careful attention to layout and wiring." A promotional brochure from Collins Radio appealed to hobby radio operators' technical aesthetic by displaying the interior of a receiver with the claim that "Neat wiring, complete shielding, and careful component layout contribute to superior performance and attractive appearance" (figure 3.2). In the same spirit, the 1951 Allied Radio catalog described the technical specifications of a Hallicrafters receiver—after just a brief mention of the cabinet color and size—under the heading "Handsome Styling."[28]

Hobbyists favored spacious parts layouts that facilitated the inspection and manipulation of electronics. Frequent articles in ham radio magazines described the alterations possible when manufacturers provided access to the inner workings of equipment. In modifying purchased gear, some hobbyists saw an opportunity to individualize equipment. "You can choose your own parts layout; and you can build as elaborately as you want," one article noted. Most hams wrote of making adjustments to enhance performance, update aging technology, or adapt a unit for a specialized task. A transmitter that had been "widely accepted by the amateur fraternity" before television became popular, for instance, could "be reasonably well cured of its bad habits" of causing interference with television reception by completing modifications that would result in "The Collins 310B—1953 Version." The Collins Radio Company recognized that certain improvements likely would be critical to preserving the good reputation of its equipment over a long period of ownership and sanctioned after-market tinkering with the designation of select "Factory Authorized Modifications." The author of a book of projects for home construction acknowledged the

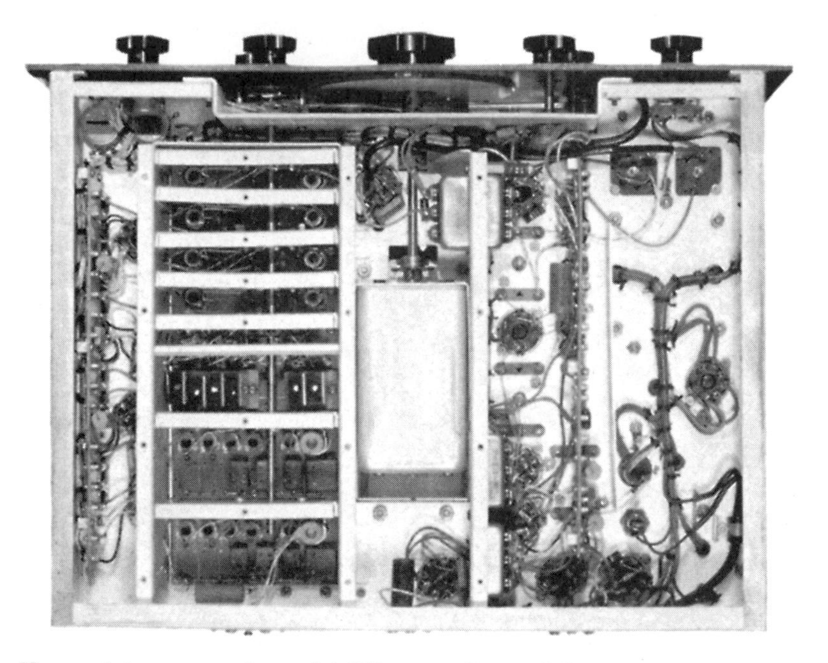

Neat wiring, complete shielding, and careful component layout contribute to superior performance and attractive appearance.

Figure 3.2
Promotional material for ham radios commonly pointed out internal features, which the tinkering hobbyist was sure to encounter. From Collins Radio Co., "The Collins 75A-1 Amateur Receiver" (1948), page 3. Reprinted courtesy of Rockwell Collins.

importance of rebuilding as a hobby activity when he singled out projects that were "particularly well suited to modification."[29] Hams evidently embarked on building their own equipment fully aware that they would later modify it.

In addition to designing equipment suited to hobbyist values and practices, manufacturers supported ham radio operators as active consumers with a continuous stream of technical lessons. Every consumer of technology needed some basic coaching. This often came through advertising or owner's manuals for specific items. Occasionally, manufacturers' technical guidance covered a broad category of products, such as the dictionary of *Common Words in Radio, Television and Electronics* put out by RCA to introduce prospective buyers to the vocabulary essential for discussing the new postwar electronics.[30] As customers who typically took their electronics purchases apart, hams needed to know far more than terminology.

Advertisements for hobby components and equipment contained data crucial for tinkerers. Asked to name their favorite part of *CQ* magazine, readers rated the ads second only to the new products section. One hands-on hobbyist reported, "Every Heath advertisement and circular was thoroughly gleaned for all bits of knowledge." To help readers quickly locate such information, *Ham Radio Magazine* provided an "Advertisers Index." Amateur radio operators treated catalogs, which were essentially compendia of advertisements, as reference books. A handbook called the large annual catalogs released by electronics distributors "veritable encyclopedias" that were "well worth having." Electronics catalogs addressed the compatibility and interchangeability of components, topics another hobby guide called indispensable for any "experimenter par-excellence." Even formal instruction in radio technology included catalogs as texts. The *Radio Receiver Laboratory Manual* contained an appendix with blank pages intended for students to record notes from the distributors' and manufacturers' catalogs made available in the classroom.[31]

Producers of ham radio equipment provided remarkably detailed technical information. The common offer of "schematics and specs" referred to schematic diagrams showing the complete electrical structure and technical specifications of construction, operation, and performance. Hammarlund Manufacturing distributed two leaflets about its HQ-120-X receiver. The shorter, four-page version covered the basics and suggested that potential customers write to request the "complete technical information." Those who did so received sixteen pages that included six schematic diagrams, four photographs of the interior, a chart of selectivity curves, and a two-page listing of all parts with descriptions and their locations on the schematics. Sections of the text explained Circuit Arrangement, Constructional Details, Operation, Realignment Procedure, Antenna Requirements, and Maintenance.[32] Though this degree of openness may have put the ideas of ham radio manufacturers at risk of being poached by competitors, transparency was the industry standard, with all players widely circulating technical details through ads and catalogs.

Project guides issued by electronics firms taught home construction at the same time as they functioned as subtle promotional materials that stimulated sales of components. Allied Radio, the largest parts dealer in the 1930s, sold "'Build-Your-Own' Wiring Diagrams" that came with parts lists to organize purchasing. Decades later, Radio Shack continued this marketing tactic. A six-volume series published on electronics building projects assured readers that "To avoid the problem of finding hard-to-get parts, each project makes exclusive use of parts and supplies available from Radio Shack."[33]

Newsletters issued by RCA and General Electric combined overt advertisements with advice on building equipment. *RCA Ham Tips*, published from 1938 through the 1970s, and *GE Ham News*, available following the war and into the 1960s, provided hobbyists with several pages of updates on products and techniques every other month. Instructions for assembly projects and features such as the *Ham Tips* contest for "an outstanding Ham rig that is 100% RCA tube equipped" encouraged readers to buy components. GE and RCA asked electronics supply stores to distribute the newsletters free of charge and, in return for this promotional service, left a space on the newsletters for the dealer to stamp its name and address. Enterprising suppliers sometimes used the materials for more narrowly targeted sales strategies. A representative from Elmar Electronics, for instance, handed out copies of the *GE Ham News* "Special DX Log Issue" at a meeting of the Northern California DX Club.[34]

Amateur radio suppliers and customers alike worried that hobbyists would be left behind when transistor technology superseded vacuum tubes in the 1950s. Since the earliest days of radio, tubes had been the fundamental components that amplified electrical signals. All hams knew how to work with vacuum tubes, and most also understood the electronics theory behind how they functioned. Hobbyists accustomed to the light that indicated a vacuum tube was working spoke of taking comfort in the "warm glow" of the familiar components. By comparison, transistors were tiny, opaque, sealed devices—black boxes, literally and figuratively—that made learning by doing nearly impossible in home workshops. This did not matter to the electronics industry, which focused on the fact that solid-state components were cheaper and more reliable than vacuum tubes.

As transistors pervaded hi-tech electronics, producers and suppliers embarked on an extensive campaign to educate hams. The first instructional literature on transistors offered technical lessons and encouragement

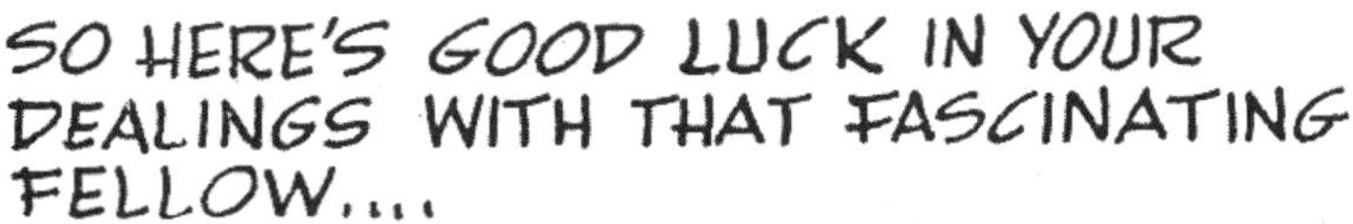

If your local jobber does not yet stock Hydro-Aire Transistors,
please write to us direct.

Figure 3.3
To ease hams through technological change, Hydro-Aire introduced the transistor in the guise of a nonthreatening cartoon character. Reprinted with permission from Hydro-Aire, *The Transistor and You* (1955), page 4.

to coax hobbyists gently through the transition from the vacuum tube to the transistor. Often manufacturers' guides, such as the pamphlet *Transistor Theory and Circuits Made Simple* that the American Electronics Company distributed in 1958, contained plans for basic construction projects to motivate transistor sales along with tinkering.[35] The hobby literature tried to make transistors seem less intimidating, repeatedly calling them "fun" and "easy" to use.

Hydro-Aire's *The Transistor and You* (1955) exemplified manufacturers' attempts to assist hobbyists through technical changes. The booklet introduced the jovial character of "Mr. Transistor," who showed tentative hobbyists the latest techniques (figure 3.3). In a "Salute to the Ham," Hydro-Aire welcomed the ham community with Morse code and flattering remarks on "the importance of the HAM fraternity, and of the great service rendered by this enthusiastic group in pioneering and development in the electronic field of endeavor." Illustrations with each building project in the book depicted Mr. Transistor playing a central role. He fired the starting pistol for a race car and operated a stopwatch on the page containing a schematic for an electronic timer, then hoisted a megaphone in front of a ham's mike on the dynamic microphone preamplifier page. The overall modernizing effects of the new components could be seen in a drawing that compared tube technology to transistor technology (figure 3.4). Transistors would relieve the frustrations of working with vacuum tubes, bring order to the ham shack, and even refine the hobbyist from a ruffian with a screwdriver tucked behind his ear into a quasi-professional—hair combed, collar buttoned, and sporting a tie—with his hand delicately turning the dial of a sleek radio.[36]

By the mid 1960s, hobbyists understood the basics of working with transistors. Efforts to enable use of the devices then included helping hams sort through the enormous variety of transistors. A transistor ratings table intended "to aid the experimenter in identifying, selecting and substituting transistor types" took up half the pages in the Capstone Electronics 1967 guide to building with the "bargain transistors" available at the time.[37] Already, though, transistors had lost their position as the state-of-the-art electronics components to integrated circuits, devices that proved to be incompatible with tinkering.

Kits—Building Commodified

The sale of accessible, ready-made equipment, with all the technical information and lessons an active user would need, made it possible for hams to be productive consumers. The sale of kits took this a step further and completely commodified building in a way that allowed hams to be consumers and producers simultaneously. Kits tamed freestyle construction and grounded the whole process in the marketplace. Without a kit, building an original design or even a published plan could resemble inventing. The electronics hobbyist had to resolve the challenges of independent building through trial and error and creative adaptation. The kit assembler, on the other hand, purchased an entire construction project in one package, with all necessary parts and specific directions for soldering them together. If questions arose, the assembler could turn to the thorough instruction manual and diagrams included with the kit, or call the kit manufacturer's customer support center. The promise of the leading electronics kit seller, "We won't let you fail," pointed out that piecing together a

Figure 3.4
A ham radio operator who accepted Hydro-Aire's friendly transistor into his shack was shown to escape frustration, toil, and general disarray. Reprinted with permission from Hydro-Aire, *The Transistor and You* (1955), page 18.

prepackaged set of components was a do-it-almost-yourself activity. Assemblers felt and appeared deeply involved with kit projects, but their production occurred in a protected, consumer environment.[38]

With respect to several key aspects—price, necessary technical understanding, time commitment, freedom of design, and available electronics lessons—kits occupied an intermediate role between buying ready-made gear and constructing from components. As the position of kits along the building-buying spectrum shifted over time, radio hobbyists reevaluated kits. A 1950 editorial in *CQ* declared kits "the biggest spur to home construction since the invention of the electric soldering iron and the chassis punch." A new editor at the same magazine offered more restrained praise in 1956, calling commercial kits "a good compromise," with some financial benefits and some of the pleasure and educational value of construction. Within less than two years, the limitations imposed by kits led him to take the harsher position that kit assembly should not count as a form of home construction.[39] This rapid change in perspective at *CQ* within the 1950s encapsulates perspectives expressed by the ham radio community across the much longer history of kits.

The radio kits available from the 1920s into the early 1940s merely consolidated component purchasing. With little ready-made equipment available and most amateurs building their own, kits simplified what could be a difficult and time-consuming hunt for parts from the still nascent network of supply houses. Allied Radio began selling amateur gear in kit form in the early 1920s. In addition to a selection of prepackaged kits, Allied was willing to "supply matched kits of parts for building any circuit described in any radio publication." Essentially, these were kits on demand. The customer just told Allied which plan he intended to follow in his leisure building project, and Allied bundled together all the necessary components.[40] The specially promoted kit products of the time also were little more than boxes of parts. Kit vendors provided only minimal instructions and rarely completed any assembly steps for buyers.

Other early radio kits were even less than boxes of parts. An amateur interested in building Allied's "All-Star Senior 7 Tube Superheterodyne" in 1935 had the choice of a "foundation kit" or an "essential parts kit." The first sold for $2.50 and included only the drilled chassis and front panel, a wiring diagram, and illustrated instructions. This option suited customers who had a good, local component dealer yet wanted to avoid the chore of forming the metal structure. To get all necessary parts delivered in one package, a hobbyist bought the second type of kit, for $31.18. Hammarlund Manufacturing Company sold similar "transmitter foundation units" in 1940 that were "hardware kits" containing only the parts too complicated to make at home or too hard to find for sale. Any standard components had to be purchased separately. Hammarlund advertised the partial kits as a solution to "the many constructional problems which have long been confronting the amateur."[41] Foundation kits allowed amateurs to skip the least desirable, mechanical steps of home building—including the troublesome metalwork of drilling neat, round holes and shaping brackets (the subject of constant complaints in the ham literature)—while remaining engaged with the radio-electronics technology that attracted them to the hobby.

Sales of ham radio kits increased after World War II when a handful of kit-specific firms and several radio manufacturers marketed complete, scripted equipment construction projects. This was part of a rapid growth in the market for all types of hobby kits, where total annual receipts increased from $44 million to $300 million in the first postwar decade.[42] The Heath Company dominated the electronics kit business soon after its entry in 1947 and until it left the business in 1992. Established in 1926 as a producer of full-scale, operational airplanes in kit form, the company went bankrupt following the death of founder Ed Heath on a test flight in 1931. Howard Anthony restructured the Heath Company a few years later as a manufacturer of aircraft parts and accessories. After its wartime contracts ended, the Heath Company bought and resold military surplus electronics components. The return to the kit business came in 1947 with Heath's creation of oscilloscope kits that incorporated surplus cathode-ray tubes. Hobbyists responded in great numbers to the first advertisement for the O-1 Oscilloscope in *Electronics* magazine, prompting Heath to introduce ham radio communications equipment in 1952. Heath's first ham radio kits, intended for customers with construction experience, contained all necessary parts and schematic diagrams but minimal instructions. Within a decade, Heath's approach changed. Instead of requiring customers to perform technical tasks on their own or learn them from a detailed manual, Heathkits arrived with more and more of the complex steps already completed. Heath advertised in 1958 that in its Mohawk RX-1 receiver "All critical wiring is done for you[,] insuring top performance." Over the course of 45 years, Heath marketed 150 types of kits specifically for amateur radio, 400 for electronics test equipment, and hundreds of other kits for general electronics hobbyists. Nearly 40,000 Heathkits for the HW-101 transceiver alone were sold—making it the bestselling ham product ever—and 1 in 5 respondents to a 1957 survey of *CQ* readers reported using some kind of Heathkit transmitter.[43]

Simplified postwar kits significantly lowered the level of technical knowledge required for assembly. While this change increased the potential customer base, it resulted not only from an effort to expand kit sales but also from the integration taking place industry-wide at the level of electronics components. Allied Radio trumpeted a printed circuit bandswitch of the early 1960s as "a Knight-Kit innovation" that "reduces assembly time to a minimum, and makes an extremely important contribution to stability and overall performance." Because the assembler had to "Simply plug the bandswitch into the RF circuit board, [and] solder," the prefabricated component guaranteed "you've made 32 error-free connections!" When Allied assured customers that "you really can't go wrong," it hit upon the very reason some hobbyists saw few of construction's benefits in kit assembly.[44]

The strict assembly programs outlined in kit instructions called for hams to systematically reproduce construction projects. Step-by-step directions, written to guide the least knowledgeable customer, subordinated amateurs' technical skills to those of professional kit designers and attempted to rein in tinkering. Kit sellers promised success to disciplined assemblers who would "follow the instructions exactly as provided" and held out the risk of failure to those tempted to stray from the precise assembly plan. "In the majority of cases," the Heath Company warned, "failure to observe basic instruction fundamentals is responsible for inability to obtain desired level of performance."[45] A space beside each step in a Heathkit manual asked the assembler to document adherence to the instructions by checking off the step upon its completion (figure 3.5).

The ham radio community incorporated kit assembly into its enculturation practices. Veteran hobbyists worried in the early 1950s that, with the coincidence of the increased popularity of the hobby and the increased availability of ready-made radios, newcomers would miss the important

Page 15

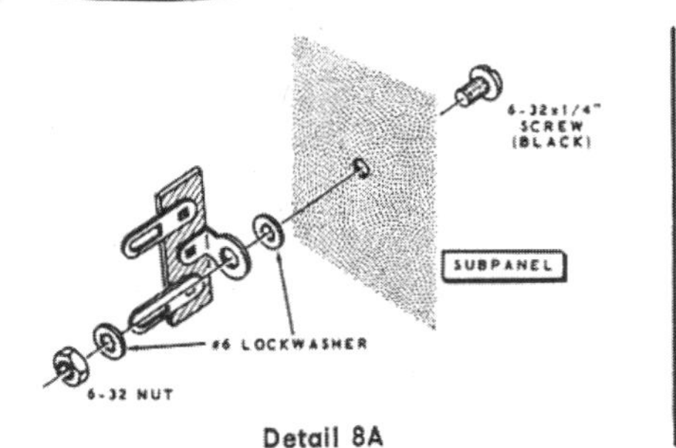

Detail 8A

Refer to Pictorial 8 (fold-out from this page) for the following steps.

(✓) Refer to Detail 8A and mount a 2-lug terminal strip at H. Use a 6-32 x 1/4" black screw, two #6 lockwashers, and a 6-32 nut.

(✓) Refer to Detail 8B and mount a 1000 Ω control (#10-158) at S. Use a control lockwasher and a control nut. Position the control lugs as shown.

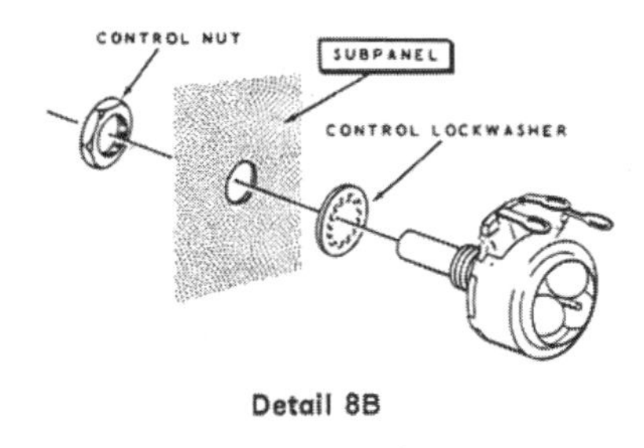

Detail 8B

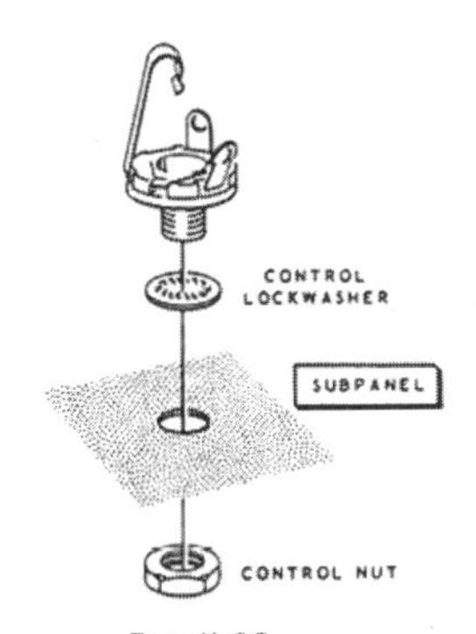

Detail 8C

(✓) Refer to Detail 8C and mount a phone jack at T. Use a control lockwasher and a control nut. Position the jack lugs as shown.

NOTE: As you mount the switches in the next three steps, place the locating tab on each switch in the small hole beside the mounting hole.

(✓) Temporarily mount the VOA Function switch (#63-431) at P with a control nut.

(✓) Temporarily mount the AC-DC Mode switch (#63-520) at R with a control nut.

(✓) Temporarily mount the MA Range switch (#63-435) at M with a control nut.

(✓) Mount a 50 KΩ dual tandem control (#12-80) at G. Secure the control by twisting each mounting tab 1/8 turn.

Figure 3.5

The disciplined kit assembler who used this manual checked off each instruction step-by-step. Reprinted with permission from Heath Co., *Assembly and Operation of the Heathkit Solid-State VOM Model IM-25* (1967), page 15.

technical and moral training that came from equipment construction. Advocates of hands-on learning hoped that assembling kits might help novices "become good technicians" by offering the same lessons in a less intimidating context. Hams additionally believed that the rigid structure of kits promoted desirable character traits. Handbooks dictated that kit projects "*must not* be hurried" and required "patience," "careful work," and "self-discipline." To achieve the proper behavior and attitude for assembly, one guide advised hobbyists to work on kits "when you feel fresh and eager, rather than tired and anxious." "Failure to successfully build a Heathkit," a hobbyist claimed, "usually revealed flaws in the kit-builder's patience and temperament, rather than the kit itself."[46] The culture of amateur radio, though, also encouraged tinkering and technical experimentation. To that end, hams supported modifications, either during or after assembly, that introduced innovation into kit assembly by resisting the tight script of the instruction manual.

Kits containing ready-made electronic elements made fitting together a kit less like building and more like purchasing a device directly. Preassembled sections reduced the time and skill required to assemble a kit, and with less to do, there were fewer opportunities for learning by doing and for the display of technical mastery. Circuits sealed inside tiny, opaque boxes prevented the assembler from even seeing what the manufacturer had provided, let alone tinkering with it. Meanwhile the relative cost advantage of kits shrank as prices for ready-made gear fell. The impact of these changes gradually accumulated over the 1950s and 1960s, then the introduction of integrated circuits into kits in the 1970s tipped the balance.

Kits played to, and for decades resolved, the tension hams felt between building and buying equipment. But the commodification of construction through kits was not an instance of the electronics industry duping customers. Hobbyists explicitly discussed ways in which kits represented a desirable trade-off. Hams could derive technical and moral lessons, cost savings, and a sense of accomplishment by putting together the parts delivered in a single package with step-by-step instructions. The factory-made cabinet hid any less-than-tidy wiring or messy soldering joints and gave the finished device the standardized look of hi-tech electronics. More than 80% of readers surveyed by *CQ* magazine in 1968 operated equipment in their shacks that had been assembled from a kit.[47] From the 1920s through the 1960s, kits successfully struck a compromise that epitomized concessions made throughout the amateur radio market to allow hams to be both consumers and producers.

Purchasing technology need not prevent technical activity. Too often casual commentators accuse producers of wresting technical control and knowledge from consumers. But the blame for black-boxed technology—or the credit for user-friendly technology, to take the other perspective—really must be shared. In response to disparate customer demands, radio manufacturers created two different styles of equipment. Receivers with simple interfaces satisfied listeners who sought news and entertainment that happened to be conveyed by radio technology. Hams, who were interested in the radio technology itself, depended on the radio-electronics industry to provide something else. Interactive products and educational support kept the hobby of ham radio alive. Whether buying parts, kits, or fully assembled equipment, radio hobbyists continued to skillfully operate technology.

4 Amateurs on the Job

In 1954, electronics technologies had revitalized the postwar economy and had so redefined technical culture as to stimulate the launch of the monthly magazine *Popular Electronics*. That same year the editor of a hobby radio magazine proclaimed, "Hams are the life-blood of the electronic industrial complex."[1] The boast contained a grain of truth. Hi-tech employers, endorsing the culture of ham radio, recruited hobbyists for the skills and traits developed through recreational tinkering. Despite hams' proud insistence at times on their status as "amateur" radio operators, there was a significant overlap between the groups that worked with electronics during the day for wages and in the evening for pleasure.

Hams had to redefine themselves at mid century. The technological wonders of electronics, spectacularly showcased in the space program and other demonstrations of Cold War technical might, displaced radio from its dominant position in the national technical culture. Playing up their connections to the powerful electronics industry reflected technical clout onto radio hobbyists but risked undermining their independence. Hams coped with this dilemma by appealing to the rhetoric of amateurism, which allowed them to claim pure motives while asserting exceptionally close ties to professionals.

The Age of Electronics

National Geographic capped rumors that had been circulating since early in World War II by reporting in 1945 that "Scientists say we are entering now upon the 'electronic age,'" calling electronics "a potent force in remaking our world." Amid frenetic war-related production and frustrating shortages on the home front, industry and consumers alike had dreamed of calmer days when technical breakthroughs would generate novel products. Enthusiasts further anticipated that electronics not only would shift patterns of consumption but would wholly reshape technical culture following World War II. RCA began distributing a free quarterly, *Electronic Age,* in 1941 to explain emerging technologies in a way that would ease the general public through the transition. Popular magazines portrayed electronic devices as strong, clean, safe, and scientific and declared that "electronics promises new miracles in industry" (see figure 4.1).[2]

To the business world, electronics offered an opportunity to increase productivity and profits. *Fortune* magazine in 1943 calculated that "From radio to radar, electronics is rocketing a $4 billion war business toward a postwar industrial revolution." Comparison to a stalwart American manufacturing sector—in the observation that the electronics industry was worth "more than the whole prewar U.S. automobile industry"—underscored that this was a "basic revolution." *Fortune* expected electronics tools to "exert a great new leverage on all industry," fundamentally altering production systems. A year later, *Business Week* cited the broad application of electronics in fields from communications and entertainment to medicine and food processing as the reason that "probably no other industry faces the postwar period with less concern about over-expanded capacity." Although the wartime boom in electronics had been enormous, the electronics industry saw "no ceiling on the postwar demand for its product." A series of four articles in *Popular Science Monthly* emphasized the versatility of electronics by saying it was not merely "an industry in itself," but rather "a technique, a way of doing things in a lot of industries." Electronics made factory production, for instance, "more and more automatic by amazingly accurate methods of measurement and control."[3]

The popular press indicated that electronics also would revolutionize everyday life after the war. In the first half of 1943 the electronics industry spent millions of dollars on advertising. *Fortune* magazine blamed the "strange, futuristic pictures of the coming age of electronics" presented in these promotions for the fact that electronics became "a glamour word—more dazzling than informative." Yet this same critique could have been leveled at numerous articles that claimed to provide straightforward

Figure 4.1
In the early 1940s, electronics held potential as a "mystery weapon" that would radically change everyday life. An article in *Better Homes and Gardens* predicted that new technologies available in the "exciting, surprising, slightly mad" years ahead might include ultraviolet emitting tubes to eliminate bacteria from classrooms, spools of wire capable of playing four hours worth of audio entertainment, and projectors to show television on large, flat screens. Illustration and text from Walter Adams, "Mystery Weapon Today, Your Servant Tomorrow," *Better Homes and Gardens*, August 1943, page 20.

information on electronics. In one such piece, John Sasso offered to educate readers of *House Beautiful* about electronics, a term which he said, "sound[ed] like something out of a 25th-century comic strip." Though in 1943 the word electronics might seem "as outlandish as 'telephone' was to your grandfather," Sasso comforted his largely female audience with the assurance that "After the war, even toddlers will know all about it." The answer Sasso provided to "What's All This About Electronics?" however, did not go much beyond vague, science-fictional prophecy. He stated that "our whole mode of living will be changed" as a result of "superautomatic control." Sasso dismissed vacuum tubes as "too complex to discuss here" and settled on describing them "as the *eyes*, *fingers*, and *ears* of tomorrow that will endow mechanical devices with human attributes." A few months later a similar article in *Better Homes and Gardens* called electronics a "Mystery Weapon Today, Your Servant Tomorrow."[4]

As manufacturers reverted from supplying the military with battlefield technologies to selling consumer goods, the predicted age of electronics arrived. This prominent technical culture forced ham radio of the 1940s to 1970s into a difficult position. In the previous three decades, sometimes referred to as the radio age, hams had stood at the forefront of technical developments. Given the Federal Communications Commission (FCC) ban on hobby radio during World War II and the emergence of a new hi-tech standard, an electronics executive in 1942 figured that hams would abandon radio operation for "electronics gadgeteering."[5] Had hams only been attracted to the latest technology, this might have been a logical assumption. But hobbyists continued to enjoy radio communication. The challenge was for radio hobbyists to maintain their reputation as technical masters in the age of electronics.

Electronics Work and Leisure

The booming electronics industry readily absorbed the returning veterans who had received electronics and radio training during World War II. *Business Week* had cited the availability of such experienced potential employees as one reason to expect peacetime success for the "electronics era." As their period of military service drew to a close, thousands of soldiers wrote letters to radio manufacturers seeking civilian work, and *Radio News* published a guide to the various available jobs and tips on how to secure one. A few cautious postwar forecasters expressed concern that there might be an oversupply of workers with electronics skills. The U.S. Department of Labor's 1948 *Occupational Outlook Handbook* called electrical engineering an "Expanding field; [with] good prospects for those already well trained," but cautioned that the job market in electronics would be tight. Those who had received a technical education in the military soon would be joined in the search for employment by graduates of engineering schools, where enrollment in the late 1940s was "more than three times as high as average prewar enrollment," with an "exceedingly high" number of students in electrical engineering.[6]

The electronics industry entered a sustained period of growth, not just a brief surge. The number of Americans working in electronics manufacturing tripled from 1950 to 1960, reaching nearly 780,000. In the early 1960s the Department of Labor predicted further expansion, to "nearly 1.1 million by 1970."[7] Any fear that there would be too many skilled workers was set aside for years. A 1960 guide to electronics careers reported that "This high paying job market is actually crying for trained personnel," with great demand for "engineers, technicians, and technical writers." Not until the early 1970s did the number of electronics workers exceed the positions available in industry. This followed from extreme cutbacks within aerospace and related fields and from the reduction of federal research funds to universities. *Business Week* declared the larger technical unemployment problem, which had left 50,000 "brains" idle in 1971, the "worst job crisis in more than a decade" for scientists and engineers. Just a few years later, though, *Mechanix Illustrated* offered an optimistic "yes" to the question posed by its article "Is Electronics Still a Good Career?" Once again, "partly because of this overreaction [to the weak job market] and partly because of industry growth, demand for engineers and nondegree techs is outstripping the supply." The Department of Labor predicted roughly 12,000 openings for electronics technicians and 11,000 for electrical engineers would be created annually until the early 1980s. As the 1980s began, jobs in electronics remained plentiful, buoyed by the expanding computer sector.[8]

The electronics industry and electronics hobbies supported each other. During the age of electronics, tinkering with resistors and capacitors was, according to a 1959 how-to handbook, "one of the fastest growing hobbies in the world" and "also one of the most exciting." Electronics hobbies functioned as a peacetime equivalent to military training in terms of offering experience that employers valued. In the opinion of the Department of Labor, "Some of the best [electronics] technicians are self-taught radio amateurs who acquired both the theoretical and practical aspects through home study and experimentation." Guides to leisure electronics frequently repeated the claim that "Many electronic technicians received their basic training at their home workbenches" and gave readers "pointed information dealing with the place of the radio amateur in the military services and with electronics as a lifetime career." A chapter on "Electronics as a Career" in one handbook named "hams who [had] made good in a big way"—including six who became presidents of radio or electronics companies—as a source of inspiration and promoted the electronics industry as "offer[ing] almost unlimited opportunities to serious, diligent workers."[9]

Occupational surveys of ham radio operators consistently found that, on average, two of every five worked in electronics. Still more had technical careers, broadly defined. The employment of the members of a radio club in Rochester, New York, in 1950 reflected this pattern. Of the thirty-three club members, nine had jobs in radio broadcasting and ten others were engineers. A physician and a farm manager belonged to the club, but most members worked for local technical companies. Eastman Kodak alone employed one third of the club's members. Others worked for Rochester Gas and Electric, Bausch and Lomb, the General Railway Signal Company, and Stromberg-Carlson, a telephone and audio equipment manufacturer. Around this time, 361 licensed hams lived within a twenty mile radius of Rochester. Sharing many individuals who transferred knowledge back and forth between shop floor and home workbench strengthened the technical firms and hobbyist communities of a region. In the early 1960s, when the electronics industry employed forty percent of all manufacturing workers in Orange County, California, there were thirty radio clubs in that area.[10]

Even in the 1960s when ham radio required less electronics knowledge, because of easier licensing tests and the greater availability of ready-made equipment, the link between technical employment and leisure held strong. The editor of *CQ* magazine, perhaps in an effort to attract diverse advertisers, asserted in 1963 that "Unlike years past, when amateurs were almost always connected with the electronics industry, today's amateurs come from all walks of life." But the magazine's own reader surveys contradicted his undocumented statement: results in 1968 duplicated those from a decade earlier, which had shown "that over 40% of our readers are working in electronics." A broader poll of ham radio operators firmly established the overlap of the hobby with technical careers. The Stanford Research Institute reported that half of licensed hams worked in radio communications or electrical engineering. More striking were the data that a total of 73% of hobbyists, compared to only 2% of all Americans of working age, held jobs in engineering and science.[11]

Participation in amateur radio served as a route to skilled employment with financial rewards. A 1960 guide to *Jobs and Careers in Electronics* sketched a typical trajectory for someone entering the position of "junior electronic technician" with only "interest and proven or indicated aptitude—a budding radio ham, for example, or a man who has tinkered with small electronic construction projects at home." "Under conditions of normal industry growth" and "with suitable training," the expectation was that "he could rise steadily" from performing "simple wiring and testing under supervision, at a salary of $55 to $70 a week," and after as little as three years become a "'senior electronic technician' at pay up to $130 a week." Though a significant gap in status separated technicians and engineers, spare time spent on technical projects also eased entry into engineering and, thereby, into the professional class. Sociological studies of the 1950s and 1960s noted that, of the major professions, engineering attracted "the highest proportion of practitioners from working class origins" and had "become an avenue of upward mobility for the intelligent sons of working-class families."[12]

The combination of the cost barrier to ham radio and the hobby's potential to boost earnings produced a group with an average income substantially above that of the wider population's. Among respondents to a 1957 *CQ* survey, the most common annual salary range was $5,000–6,000. This roughly corresponded to the mean income of Americans in the second highest of five wage-earning categories established by the Bureau of the Census. The average hobbyist salary reached $7,350, and one fifth of readers made more than $10,000 a year. In the mid 1960s, hams received a median income of "about $9,900," more than twice that of male workers overall in the United States.[13]

ENGINEERS and SPECIALISTS alike . . . if you are qualified by experience or training in the design, maintenance and instruction of Communication, Radar and Sonar Equipment — Philco NEEDS YOU NOW! The assignment: a wide range of commercial and government operations to service on a long range basis.

As the world-pioneer in servicing electronic equipment, UNLIMITED OPPORTUNITY and JOB SECURITY are more than just "sales talk" . . . in addition to TOP COMPENSATION and special assignment bonuses, PHILCO'S many valued benefits include hospitalization, group insurance, profit sharing, retirement benefits, merit and faithful service salary increases.

From the perspective of technical firms, radio hobbyists made ideal employees. Hams arrived on the job with electronics skills developed through leisure hours spent tinkering and needed less training than other new workers. Just as important, a hobbyist's identification with technology inspired a personal closeness to the business. After-hours tinkering, which provided self education, was taken as a sign of employees' dedication. "A large proportion of the [7,000] engineers" from General Electric surveyed in the late 1950s "noted the importance of leisure time interest in engineering" to advancement in the technical workforce.[14] Electronics companies knew that hams would spend occasional evenings on activities related, if not directly applicable, to daytime work, without the incentive of overtime pay.

Technical employers welcomed the culture of ham radio into the workplace. To find out "What Industry Thinks of Ham Radio," the author of a 1960 handbook spoke with executives and managers at General Electric, RCA, and Collins Radio. Each mentioned the value of hams' experience and practical electronics knowledge, then went on to point out that certain personality traits associated with hobbyists were critical to an electronics career. The manager of GE's Engineering, Radio and Television Department stated that radio enthusiasts had a "confidence" with technical matters, "which means a definite edge over [other] fellows." Even long after active involvement with radio, former hobbyists remained favored employees because they were "still very much hams in spirit." Not surprisingly, hams frequently repeated the praise that skills and attitudes developed in technical recreation translated into success on the job. At the beginning of the 1980s, the author of a book on electronics projects claimed that hams' "tinkering and improvising abilities" continued to lead them to "achieve results above those of their co-workers."[15]

Electronics manufacturers especially, but other technical companies as well, sought to increase the number of hams on the payroll. A radio magazine's 1953 article on the "Hams in Industry" noted that 125 hobbyists

Figure 4.2
Hi-tech firms recruited skilled employees through ham radio magazines. This call by Philco welcomed applications from anyone "qualified by experience or training." Philco advertisement, *QST*, August 1955, page 97.

worked in the Research and Development Laboratories at the Hughes Aircraft Company in Culver City, California. Little more than a year later, Hughes was one of many firms running help-wanted advertisements in the same hobby magazine for "electrical engineers or physics graduates with experience in Radar or electronics or those desiring to enter those areas." Employers also placed job listings in the newsletters of local radio clubs, a recruiting strategy that was both cheaper and precisely targeted geographically.[16]

Producers of ham radio equipment did not need to bother advertising jobs to hams. The Collins Radio Company explained that the many hobbyists on its staff "had been attracted by the reputation of Collins equipment." The employment of hams functioned to attract even more as customers. Radio manufacturers frequently noted in advertisements that hobbyists built the equipment. Collins, for instance, described one of its transmitters as "designed by engineers to whom CQ is a cherished and friendly sound."[17]

Many technical firms actively fostered recreational radio. Simple gestures made by a company showed an appreciation of hams and, in some cases, developed into a further commitment. When hobbyists at GE expressed an interest in forming a directory to locate each other, for example, a company newsletter issued the call for participants. Later GE joined a long list of companies—Kodak, RCA, AT&T, CBS, Eastern Air Lines, and Lockheed, to name a few—that supported employee groups that met during lunch breaks or after work to talk about ham radio. The Lockheed ham club came under the umbrella of the Lockheed Employees' Recreation Association, a program designed to enhance workers' quality of life. In the late 1970s, the Recreation Association at the Burbank, California, plant included more than thirty specialty leisure groups. Lockheed granted the radio club space for a station and for holding meetings, a modest monthly budget, and access to a kitchen, classroom, and auditorium shared by the other clubs. If the hams faced a major expense, such as the purchase of new equipment, they could apply for special funding from Lockheed and routinely received it.[18]

Hi-tech businesses also lent assistance to independent radio clubs that had only a few of their employees as members. The Rochester Amateur Radio Association (RARA) declared in 1957 that it was "very indebted to many industrial organizations in Rochester for their loyal and unquestioning support of our various activities." Long before Kodak began to sponsor a radio club for its own workers in 1966, the company "contribute[d] support in many and various wonderful ways" to the local radio club. In 1950, for example, Kodak gave RARA an "entire room" for the club's display at the second annual Kodak Hobby Show. RARA's newsletter called Rochester's General Railway Signal Company "one of our best friends" for fully financing the truck the club used for mobile emergency communication drills, including the expenditure of nearly $200 to prepare the vehicle for its 1957 state inspection. Since it was "doubtful that the Emergency Truck provides advertising that would bring in any new orders for G.R.S.'s highly specialized items," RARA deemed the company's contributions purely altruistic and an example of "why Rochester is such a fine town in which to live and work."[19]

The cultural clout attached to electronics offered hams rewards less tangible than paychecks and sponsorship of club activities but very real all the same. Through their technical employers, assemblers and engineers alike became associated with progress, control, and national security. "This is a pushbutton age," wrote one proponent of ham radio in 1957. "America's survival in a scientific era may well depend on its technicians."[20] Ten years later, a vocational school lured students with the prospect of earning respect in electronics careers. "Behind today's microwave towers, pushbutton phones, computers, mobile radios, television equipment, guided missiles, etc.," began the Cleveland Institute of Electronics advertisement, "stand THE TROUBLESHOOTERS—the men who inspect, install, and service these modern miracles." The accompanying drawing of a man standing above a landscape dotted with sophisticated devices reinforced the idea that those who mastered electronics became strong, masculine protectors (figure 4.3).[21]

Hams publicly identified with leading industries to benefit from the excitement surrounding innovation in the age of electronics. The warm glow of radio tubes lacked luster alongside sleek transistors and integrated circuits and the compact devices built from these hi-tech components. To maintain an appearance of technical leadership, hobbyists emphasized workplace achievements. The separation of ham radio technology from the cutting edge seemed most obvious, and the connections hams professed to the contrary seemed most contrived, with regard to the space

Figure 4.3
An illustration in a vocational school advertisement enticed readers to "Join THE TROUBLESHOOTERS" and portrayed an electronics technician as the master of a wide range of "modern miracles." Illustration and text from Cleveland Institute of Electronics advertisement, *CQ*, October 1967, page 32. Reprinted with permission.

race. When news of the manned NASA missions of the 1960s captured Americans' attention and imagination, hobbyists played up their ties to the aerospace industry. The federal government deliberately had constructed a technical identity for space flight and the astronauts on board, producing what Michael Smith later called a "display value" that "equated technological preeminence with military, ideological, and cultural supremacy."[22] Following on the government's success with fixing a positive image of NASA in the public's mind, hobbyists attempted to bask in reflected glory.

Without question, a number of hobbyists directly contributed to the space program. The path of Bob Murphy shows how technical training gained through ham radio and military service could lead to a career in electronics connected to NASA. After three years as a World War II radio operator, Murphy attended college, then took a job in 1950 with an electronics manufacturer in Palo Alto, California. A decade later his work routinely included business trips to Houston and the Mercury Control Center at Cape Canaveral. The hectic pace of the aerospace industry in the early 1960s forced Murphy to curtail hobby activities, resign his post as editor of the Northern California DX Club's newsletter, and eventually to move to what he referred to as "W5 land" (based on the FCC's distribution of license numbers according to geographical areas) to be closer to the Manned Spacecraft Center in Houston.[23]

Hobby publications that suggested that hams in general were instrumental to the space race overstated their significance. *CQ* ran a photograph of the one astronaut in training for the Apollo Program who held a ham radio license on the cover of the magazine in 1965. Inside, the editorial interpreted this as an example of how "Almost daily we receive small indications that amateur radio is getting more intimately involved in the space age." Four years later, in anticipation of the moon landing, *CQ* encouraged all hobbyists to feel proud "of the role played by so many of our fellow amateurs in this staggering achievement." The magazine, citing "amateur industry sources," claimed there were almost 10,000 hams "whose jobs involve them in the aero-space industry and who are therefore entitled to puff out their chests with a little extra pride at being once again in the front rank of scientific development."[24] Given the complex and varied systems necessary to NASA operations, it may very well have been true that so many hobbyists worked for firms that in some way supported the aerospace industry. But to credit each, while stressing his identity as an amateur, was a rhetorical turn intended to let hams share in the prevailing technical enthusiasm.

Electronics hobbyists and firms mutually benefited from the connection of home tinkering to work practice. Courted by employers with preferential treatment, ham radio operators found their leisure pursuits appreciated and their image as technical masters validated. Electronics companies in return profited from hobbyists' skills and devotion to technological matters. Only a hobbyist would have gone so far as to call hams "the life-blood of the electronic industrial complex." Technical firms that recruited hobbyists tacitly agreed that radio amateurs did supply some vital element to the realm of professional electronics.

Invoking an Amateur Identity

While enjoying many advantages from their association with the electronics industry, hobbyists proudly adopted the label "amateur" to stake out a certain independence. The term "amateur radio operator" is confusing, possibly intentionally so. Radio amateurs never existed in a tidy dichotomy with a particular group of professionals. Hams during the radio age described themselves in relationship to professionals in a vague sense, claiming contributions to the emerging field of radio as significant as those made by professionals, for instance, and modeling hobby organizations on professional associations. But largely this just reflected the early twentieth century buzz about professionalization.[25] Radio broadcasting differed from hobby radio in the sense that broadcasters only transmitted and hams participated in two-way communications. Though wireless communication in the military or on board commercial ships came closer to being a paid version of ham radio, the hobby encompassed many activities beyond establishing person-to-person contacts. To satisfy the cultural norms of the ham community, hobbyists also had to perform construction, repair, or modification tasks that resembled the work of radio engineers and electronics technicians. Literally, then, hams were not amateur radio operators—there was no clear professional referent—and the great number of hams employed in electronics were not amateurs in relation to that technical domain either.

Instead of suggesting a precise counterpart to a specific radio occupation, the "amateur" in "amateur radio" classified hams' pursuits as wholesome. The name amateur radio displayed the tension in the dual technical identity composed by radio hobbyists. It subverted the traditional sense of amateur and instead signified an intermediate category split between workplace and leisure realms.[26] The common definition of an amateur as someone motivated only by the love of an activity connoted integrity. Few hams directly benefited financially from their hobby—in fact, for most it proved an expensive indulgence—but given the strong connection between basement tinkering and successful industrial careers, it seems disingenuous to overlook motivations other than blind devotion to radio.

The language of amateurism suited the image that hobbyists sought to create through their technical identity. Radio enthusiasts wanted to represent their technical interests as stemming from pure inventive and exploratory goals. The categorization as amateur distinguished leisure from work, private commitments from employers' profit concerns. Additionally, isolating a separate mode of electronics practice reasserted hams' control over entire projects and emphasized hands-on skills in a way that alleviated frustrations workers felt with managerial oversight, automation, and the division of labor. Related workplace complaints were expressed at mid century through works such as *White Collar* (1951) and *The Organization Man* (1956), which warned that corporate management eroded men's sense of self.[27]

Hams' complicated hybrid identity as amateurs and professionals was one element of the distinct technical culture they crafted. Hobbyists publicly promoted ties to the electronics industry to enhance their reputation for technical mastery. On the job, hams invoked the amateur persona. The particular styles of technical knowledge and practice associated with amateurs, hobbyists claimed, carried over into paid occupations. By this logic, professional success stemmed from amateur status, completely contradicting the usual meaning of amateur.

Like other amateurs and technical hobbyists, hams embraced learning by doing as the most thorough form of education and as ideally suited to self education at the leisure workbench. By teaching mind and body together, practical training was said to cultivate a "working knowledge" beyond what could be learned in books. The postwar emphasis on formal scientific instruction for engineers contrasted with hobbyists' preference for do-it-yourself lessons. Advocacy of tinkering as opposed to research and design allied the amateur and professional electronics communities with separate traditions of practice.

To hobbyists, tinkering meant technical interactivity and a willingness to break from standard operating procedures when necessary. They felt this empirical style produced tried-and-true results. The editor of a compilation of hints that hams had shared through a magazine column characterized the entries as "generally not the products of textbooks and slide rules, but rather, the products of experimentation, experience and an almost intuitive feel for the art and science of amateur radio." With tinkering as the means, the end was "practical, bench-tested kinks and ideas gathered from the stations and workshops of hundreds of amateurs." The lessons gained through tinkering accumulated as instinctual approaches to electronics problems. Inside the electronics industry, hams bragged of applying their "exceptional amount of *working* electronic knowledge [. . .] as needed (and in their own way!)."[28]

A preference for tinkering or for formal learning was often linked to socioeconomic class and incorporated normative judgments. Since the late nineteenth century, engineers had debated the virtues of learning on the shop floor versus in the classroom. This of course had implications for who had access to a technical education, since it might come for free or at a price.[29] Association of study with the wealthy and of tinkering with the working class hung on into the age of electronics. A 1961 *Saturday Review* editorial on amateur science juxtaposed "well-to-do amateurs, men of leisure and education who studied nature purely for the love of it" with "the pragmatic workshop tinkerer, who lacked formal education but who often won through to important insights simply by 'monkeying around' with the materials of nature."[30]

Long after the general reorientation of American engineering education to a science-heavy curriculum in the 1950s, proponents of hands-on learning remained. Eugene Ferguson, schooled as an engineer in the days when practical training was customary, proclaimed in his 1992 history of engineering practice that there had been a postwar drop-off in creativity and design skills. "Deep knowledge" also diminished, according to Ferguson, and technical professionals less commonly possessed "comprehensive understanding and appreciation of the many, many facets of a situation." Taking issue with an engineering text on expert systems that valued theory (and derided "experiential, superficial knowledge"), Ferguson insisted that a distaste for physical closeness to technology "epitomizes the gulf that will always exist between expert systems and experts."[31] Technical firms signaled an appreciation of educational tinkering similar to that articulated by Ferguson when they sought to hire hobbyists.

Ham radio operators pointed to their methods for transferring knowledge, like to those for acquiring it, as evidence that the hobby was accessible to anyone with technical ability or ambitions. Rather than guarding information the way industry did with patenting and corporate secrecy, the ham community exchanged information openly among practitioners. Hobbyists feigned ignorance about the potential benefits of intellectual property law, such as when the author of a 1962 article called it "probable that many amateurs are not aware of the nature of the Patent System in the United States as a means of protection for invention." To meet technical challenges, ham culture dictated freely sharing workbench wisdom. The American Radio Relay League's code of conduct listed "friendly advice and counsel to the beginner, kindly assistance, cooperation" as "marks of the amateur spirit." Hundreds of local clubs facilitated the informal exchange of ideas among hams. The promoted "feature" of a 1949 meeting of the Rochester Amateur Radio Association was "a technical problem discussion in which a group of radio-engineer hams will answer questions and discuss problems presented by the audience."[32]

Hams used multiple channels to share electronics know-how. For one, hobbyists passed printed reference materials through preestablished social networks. "Just to show how these technical manuals circulate around among the hams," the *Schenectady Amateur Radio Association Newsletter* mentioned that "Jeffrey is wondering when Dal Hurd will return the Antenna Handbook that Jeff loaned Ted Swartz." Ham radio magazines and club newsletters provided a timely forum for swapping technical information. The nationally distributed periodicals published regular columns with tips sent in by hobbyists. Editors encouraged casual submissions, in hams' characteristically plain language. When *CQ* began running "Inside the Shack and Workshop" in 1947, it told readers, "Don't worry about literary form—just get your ideas down on paper and include rough sketches, diagrams or photos if you have them." The presentation of members' ideas in club newsletters adjusted to the amount of material available in a given printing cycle. In busy months, *The RaRa Rag* contained both a "Technical Topics" column and a separate listing of "Hints," while at other times issues ran only one or neither of these sections. Some submissions were unattributed; some named an author, if only by call sign.[33] The pages of hobby publications allowed for genuine interchange by printing feedback in response to earlier suggestions.[34] The communications tool central to ham radio also provided an important conduit. "The never-ending exchange of technical information between hams" constituted "the more serious side" of the hobby as presented in *Woman's Day* magazine. "When one amateur discovers a method of improving his transmitter by the inclusion of some new gadget he's dreamed up," a 1950 article reported, "he will spread the news far and wide."[35]

Because on-air discussions were subject to unpredictable transmission conditions and could not convey critical visual information like wiring plans, even hams who spoke regularly by radio sent letters to supplement conversations. "Please let me know how you make out by mail," Arthur Ericson wrote to his friend Andy Shafer in the early 1970s, explaining that their radio connection was "inadequate for good QSO [communication] due to the skip." The two hobbyists were trying to replicate an early twentieth century receiver. Ericson had sent the device to Shafer, but its detector—a glass tube filled with iron filings—was proving "very tricky, stubborn." "You have to have patience with them," cautioned Ericson. The detector needed fine tuning, and Ericson provided detailed instructions, illustrated with schematics, that walked Shafer through every step and warned about common mistakes. After following the advice in Ericson's first two letters, Shafer reluctantly reported, "I am sorry to say I was not able to tune in any station." Ericson responded with further information and an additional diagram.[36] Their correspondence exemplified the free exchange of knowledge that hams applauded. Naming openness as an "amateur" trait was a subtle way of distinguishing ham radio values from industrial electronics culture.

Ham radio operators imbued their technical identity with a selective form of amateurism to facilitate passage between recreation and work worlds. They exploited the ambiguity in their status, offering no fixed answer on whether an individual who supervised tube manufacturing at General Electric during the day and tinkered with those tubes for radio communication by night should be called a professional or an amateur. Creating an intermediary position for amateur radio afforded hams the advantages of identifying with technology and the freedom of men of leisure. When seeking credit for contributions to hi-tech developments, hobbyists stressed connections to professional electronics. When seeking independence, they stressed amateur qualities.

This fragmented technical identity improved the images of people and machines. Emphasizing the hobby side of radio tinkering served to normalize hams' behavior. With more than half of "amateurs" employed in technical fields, their activities could have been mistaken for an obsession with work. A sociologist in the late 1960s assessed engineers as "narrow of interest" and "relatively uninterested in 'cultural' things." According to this description, engineers failed to fulfill the middle class obligation to participate in leisure, defined in opposition to work.[37] To designate after-hours electronics as an amateur or hobby pursuit drew a line of critical social importance, separating workplace from home, corporation from self. Leisure electronics practice, though, functioned as a form of occupational training, and hams switched amateur for professional alliances whenever it suited their purposes.

The identity hams created for radio technology similarly split between amateur and professional worlds. On the one hand, "amateur" functioned as a veil of modesty for radio equipment. Powerful machines lost a bit of their threatening edge in the hobby context and seemed more like toys. This designation was especially important when Cold War anxieties provoked suspicions of international two-way radio communication being used for clandestine activities. At the same time, however, hobbyists claimed a close connection for radio to achievements of modern electronics that were popular with the public. Acknowledging hams' contributions to the space race affiliated ham radio with a technology perceived as heroic, but not dangerous. When they moved from the hi-tech workplace into home radio shacks, hobbyists transferred some of the glory of modern electronics onto older radio technology.

5 Hobby Radio Embattled

Two-way radio provided strategic communications during wartime. The military used wireless technology to contact distant troops, to listen in on enemy plans, and to spread propaganda. In World War I, radios still were rather tricky, unpredictable devices, and there was an inadequate supply of equipment and personnel with the skills to operate it. A quarter century later, the press dubbed World War II "the radio war" and radio operators its heroes. "In modern war," *This Week* magazine explained, "radio barks the commands." *Liberty* magazine credited the licensed ham with being "the guy who has won the radio war." Radio was so essential to security that the state temporarily cut off recreational access to the airwaves. The Federal Communications Commission (FCC) banned two-way radio as a home-front hobby when radio entered service as a battlefront tool in the World Wars.[1]

Amazingly, these were the only shutdowns of amateur radio. Despite dramatically publicized risks of open, international communication, the FCC allowed hams to remain on the air throughout the Cold War and the conflicts in Korea and Vietnam. Hobbyists maintained operating privileges by playing to Cold War fears. Their ongoing public relations campaigns stressed the value of keeping skilled radio technicians in reserve for the military and presented radio as a form of backup communication to be used if an attack or natural disaster knocked out regular systems. The result was an intermediate, ambiguous position for hams, similar to the relationship created as "amateurs" with regard to the electronics industry. As radio operators for civil defense, hams appeared powerful and patriotic, serving the state in an innocuous civilian role.

Defending Radio

The silencing of radio as a hobby followed directly from its usefulness as a war technology. The U.S. armed forces recruited self-taught hams and swiftly converted them into military radio specialists throughout the twentieth century. During World War I, the enlistment of hobbyists was critical because almost no one else had experience with radio transmission. In 1917 alone, 4,000 licensed amateurs joined the Navy or the Army Signal Corps, and many others donated equipment from their home stations. Radio-electronics skills only increased in importance to the military over the next fifty years, such that hams could meet only a fraction of the tremendous need. *Business Week* estimated that a staggering two million individuals were trained as radio technicians during World War II. To teach a non-ham to be a military radio operator in 1941 took three to four months. For hobbyists, the process could be cut to a mere two weeks. The Navy considered its radio technician course "one of the longest and most rigorous of the naval programs for enlisted personnel." And, until demand outstripped the supply of qualified applicants, that course accepted only those with "some previous experience in radio and electricity." Into the 1960s, the military offered ham radio operators recruiting bonuses and small incentives like ensuring that hobby shops on board ships and in bases carried ham radio supplies. The Naval Reserve, for instance, invited hobbyists in 1964 to take an examination instead of advancing slowly through the ranks to reach the status of radioman at the pay grade of petty officer.[2]

Ham radio groups encouraged members to volunteer for military duty and subsequently argued to keep federally granted frequencies based on hobbyists' patriotic service. Alliance with the armed forces proved the value of hams' technical mastery and linked them to the physically strong, masculine image of soldiers. Associating hobbyists with the military was part of the public relations campaign the American Radio Relay League (ARRL) continuously and explicitly waged in order to make outsiders "understand and appreciate our hobby." The ARRL, founded in 1914 ostensibly as a club, took upon itself the role of official, national ham radio advocacy group. Hobbyists over the years occasionally complained that the League (based in Hartford, Connecticut) was, in the words of a California ham, "a regional organization that represents a mere thirty percent of the licensed amateurs in the U.S." and could not speak for the entire hobby. Support for the ARRL peaked when the hobby was in jeopardy. During the wartime shutdown of recreational radio, membership increased by a third and included a very high percentage of licensed hams. Membership then tapered off through the 1950s, after which the League never again included a majority of licensees. Although the editor of a Rochester, New York, club newsletter acknowledged that there were "quite a few amateurs who are dissatisfied with, or are completely indifferent to ARRL," he pointed out that the "ARRL IS THE ONLY REPRESENTATIVE WE HAVE EVER HAD DEFENDING OUR FREQUENCIES at the FCC and at these international conferences." Numerous remarks in hobby publications followed this pattern, expressing frustration with the League but conceding that it fulfilled a critical function by employing public relations and lobbying personnel to fight for the hobby.[3]

"One of the League's jobs is to maintain a public opinion generally favorable toward amateur radio," explained a 1941 article in the ARRL's magazine, *QST*. The League's efforts included national advertising campaigns, appeals to legislators and regulators, and coaching hams in techniques for promoting a positive image of radio. An editorial in *QST* called the "bread-and-butter publicity, obtained by alert radio amateurs and clubs [. . .] the backbone of the ARRL publicity program." To relieve hobbyists of creative burdens and ensure that they sought "the right kind" of attention, the ARRL provided sample documents that "may be altered to meet your local needs." Such templates ranged from a 1922 memo with "some honest propaganda on the amateur's position in the radio art" to a 1960 booklet, "Getting Newspaper Publicity for Your Club and Amateur Radio," filled with speeches and press releases.[4]

When American involvement in World War II seemed imminent, hobbyists tried to avert a repeat of the ban the FCC had placed on recreational radio during World War I. Ham organizations, bluntly discussing their motivations, rallied members to defense duties as a tactic to sustain hobby radio. A 1941 equipment catalog proclaimed, "Every serious Amateur these days is thinking in terms of *preparedness*—not only for National Defense, but for service to his community and the future of Amateur Radio." One group of hobbyists in the San Francisco area formed the Amateur Radio Defense Association in the fall of 1940 to unify preemptive efforts to stay on

THIS MAN MUST REMAIN ON THE AIR

"He Also Serves"

A Tribute to the Radio Amateur

Unseen, unsung, "he also serves"
And from his duty never swerves!
Alert to ev'ry call for aid,
Dependable, and unafraid!

Regardless of reward or gain,
He reaches out o'er hill and plain,
To render aid where e'er he can,
Or chat a bit, like man to man.

On living, vibrant ether wave
He modulates his call, to save
The victims of disaster grim,
Who face the loss of life or limb.

His call "gets through" when others fail
To rouse the help for those who ail.
Through all the day and weary night
He guards against Fate's arrow flight.

With key or mike this "minute man"
Forever gives the best he can,
And only asks that we maintain
His services to lighten pain.

Must he, a patriot, proved and true,
Be barred. as is that spoiler crew
Who seek to wreck our nation's weal
And make us to a victor kneel?

Nay. God forbid! that such be done
To him who fights against the Hun
Whose rattling sabre sounds the threat
To what we've earned by blood and sweat.

For if we rule that he must go,
We quench the fires which foil our foe.
Heed well his plea. his simple prayer.
That he be kept upon the air!!

the air. Here the phrase "amateur radio defense" had a double meaning: along with promoting ham radio as part of national defense, the Association wanted to defend the hobby. The first "Call to Action" explained that "If the amateurs can prove that their service is essential there is good reason to hope that their stations will not be ordered to be shut down in the event of a new war." The Association's magazine advocated "preparedness" among hams "as the first law of preservation."[5]

Amateur Radio Defense magazine projected an image of the typical hobbyist as manly protector and patriotic servant. The first issue called the mobilization of ham radio for national defense "not child's play but man's work" and prominently represented this masculinity in a page-two poem and illustration (figure 5.1). The verses of "He Also Serves" described the "unseen, unsung" radio hobbyist as a guardian "Alert to ev'ry call for aid, Dependable, and unafraid!" Although his primary quest was "To render aid where e'er he can," the poem admitted that the ham might also "chat a bit, like man to man." Next to the text stood a man as tall and strong as Paul Bunyan, talking into a portable radio while watching over a farming community. The label "The Minute Man of Radio" associated him with another American folk hero, the ever-ready militia fighter of the Revolutionary War. With the goal "to tell the cockeyed world, now and all the time, that amateur radio has always been, is now, and always will be doing its full part in perpetuating 'the American way,'" *Amateur Radio Defense* magazine painted hobbyists as "a reserve army of fifty thousand licensed amateur radio operators who stand ready and able to meet any threatening disaster."[6]

The American Radio Relay League stepped up its usual public relations efforts in an attempt to prevent the wartime shutdown anticipated in 1940. While ARRL leaders kept "in close touch with official Washington" to draw attention to "the imperative need in the national interest to maintain amateur radio," they charged hobbyists with partial responsibility for the outcome of negotiations with regulators. The League told hobbyists to

Figure 5.1

Arguing to keep ham radio operational throughout World War II, this poem and illustration represented the hobbyist as "a patriot, proved and true" who watched over American soil and life. Arthur H. Halloran, "He Also Serves," *Amateur Radio Defense*, November 1940, page 2.

embrace opportunities to volunteer as a way to "justify the license" and "preserve our amateur radio." *QST* magazine recommended that hams make a new year's resolution "to be more useful operators to Uncle Sam (and to ourselves)" in 1941. To avoid bad publicity, the ARRL constantly admonished hobbyists to obey radio rules. "The internal situation is such that even minor infractions can hurt amateur radio like hell," scolded a June 1941 *QST* editorial in reference to the FCC having caught some violators. The League "urge[d] caution, circumspection and restraint" to keep the hobby "above suspicion."[7]

As international tensions increased in 1940, hams' free access to the airwaves and private ownership of powerful equipment seemed risky, and the FCC issued a series of new rules that tightened control of ham radio licensing and operations. The first made it illegal to hold a conversation with a foreign hobbyist. (Most countries had entirely shut down ham communications already, so a foreigner on the air likely would have been operating against the rules of his own country as well.) The next order severely limited the frequencies available for communication via mobile equipment, with exceptions granted for emergency communications and drills during weekend daylight hours, as long as the district FCC inspector was notified in advance. Then came Order 75, which hams called "difficult and annoying in the highest degree."[8]

FCC Order 75 brought hobbyists under the strictest regulation they ever faced. It required all amateur and commercial radio operators to submit information on citizenship, military service, any trips taken outside the country, and the citizenship of their close relatives. Additionally, every licensee had to provide a "passport-type photograph" and a set of fingerprints, and these identification data had to be certified by a municipal, state, or federal official. Forms mailed to each license holder were to be completed, signed under oath, and returned with documentation within two months. The experience of Henry Broughton, born in rural Illinois in 1865, demonstrates how frustrating Order 75 could be. In an attempt to locate any document or person who could attest to his citizenship, Broughton compiled a folder thick with copies of letters to and apologetic responses from the Illinois Department of Public Health, a county clerk, the church where he had been baptized, the first school he attended, and a local radio club.[9]

The FCC made sure its new regulations had teeth. Overburdened by confirming licensees' citizenship and revoking the licenses of those who "refused" to provide documentation, the Commission hired five hundred more employees. It also stepped up surveillance of hams' behavior by creating one hundred additional patrol units to monitor the airwaves. For communicating with foreign stations, nineteen hams lost their licenses in June 1941 alone.[10]

Coverage of the crackdown in the popular press damaged the reputation of ham radio. *Variety* magazine told of the FCC discovering an "increasing number of suspicious broadcasts, involving at least two cases where amateurs have led the government on a wild goose chase by pretending to be Nazi spies." In an article provocatively titled "Radio Spies Are Trapped by Direction Finders in Prowling Motor Cars," *Popular Science Monthly* claimed the FCC's network of "direction-finding units in automobiles, fixed listening posts at 200-mile intervals, and ten long-range direction-finding stations" had caught and charged more than a thousand hobbyists with operating infractions (without any mention of espionage). *Time* magazine reported that the FCC's monitors had detected and punished several hundred illegal ham transmitters, some of them "dangerous," during 1941. Given the sparse FCC data available on revoked ham licenses, and the failure of these magazine stories to cite sources, vague figures in the thousands probably overstated the number of actual infractions and should be read more as an indication of the fear that two-way radio threatened national security. A 1941 article in *Harper's Magazine* attempted to counter the negative publicity about ham radio, which it attributed to "agitation over real or fancied Fifth Column activities" conducted on the airwaves, by refocusing attention on the "very real achievements of the hams in national defense."[11]

Despite hams' extensive campaign about the service potential of radio, the FCC banned all hobby transmissions immediately following the attack on Pearl Harbor in December 1941. The risk was simply too great that rogue transmitters could send counterfeit messages to military personnel or interfere with authentic messages. The ARRL pressured the FCC to get hams back on the air in a capacity that would be viewed as productive and patriotic. Just six months later, the FCC responded by organizing the War Emergency Radio Service (WERS) to prepare amateurs to provide emergency communications. WERS activity primarily came under the authority of the Office of Civilian Defense, but the Defense Communications Board—composed of "ranking radio men" from the FCC, military, and State Department—kept watch to minimize strategic risks.[12]

QST magazine frequently had referred to WERS even before the program officially began. Since improving the public image of ham radio was a key reason to have hobbyists participate in WERS, the ARRL wanted to ensure compliance with every rule. Over the course of 1942–1943 *QST* ran eighty articles, columns, and notices regarding WERS. In 1944 the ARRL issued a manual guiding hams through the steps necessary to establish a local WERS unit. Part of the WERS application, for instance, required an explanation of the "methods used to ascertain the loyalty and integrity of radio station operating personnel." Typically hams turned to the local police department to pass this judgment. "If there is any reasonable doubt as to a participant's loyalty or integrity," the ARRL's *Manual for the War Emergency Radio Service* recommended nothing be left to chance and that "he should be investigated by the Federal Bureau of Investigation."[13]

Although WERS granted hams a civilian role with ties to the war effort, it authorized only minimal communications and offered none of the fun and freedom of recreational radio. The FCC set aside four hours per week for volunteers to perform WERS tests in the first six months of the program and after this introductory period restricted drills to just two hours a week. WERS regulations otherwise permitted stations to go on the air "only during or immediately following actual air raids, impending air raids, or other enemy military operations or acts of sabotage." The ARRL told hobbyists to appreciate this limited access as an incremental improvement over the complete shutdown and not "to be fussy over our disappointment that we don't get to operate our home stations with our own calls." Those involved in WERS, the *QST* editor assured readers, would "still be we amateurs in our other pants."[14] To continue using radio during the war, hams had to shift masculine identities, figuratively changing out of civilian pants and into military pants.

The fact that radio could be employed to either support or hinder military operations made civilian defense groups wary of radio. In 1941, a civilian defense study, *The Specter of Sabotage*, warned that "outlaw" radio operators "might attempt to spread confusion in time of emergency." The

New Jersey Defense Council acknowledged the dangerous power of radio when debating the merits of possible forms of communication. The council decided in 1942 that "the use of amateur radio should not be considered at this time in view of the restrictions imposed on such service by the Federal Communications Commission for obvious military reasons." Other state agencies followed the model of the armed forces and incorporated radio into tactical communications. Minnesota set up its own Defense Organization to replace the National Guard troops who had left for federal service. Under this plan, radio and general communication duties fell under the Division of Military Defense, rather than that of Civilian Defense. Hams became unofficial servicemen within a "military unit" that would respond "in case of any enemy action anywhere in Minnesota."[15]

Hams quickly learned the value of presenting themselves as militaristic, civilian communicators. At the start of 1941, the ARRL solicited information about the number of hams serving in the military with the hope that "such data will be of great interest in the continuing representation of the interests of amateurs." When an Army survey later that year found the majority of ham radio license holders "not eligible for active military service because of their age (average 30–31 years), marital status (60% are married), having dependents or because of their physical condition," the ARRL shifted its attitude about associating with the military. An analysis of the survey results in *QST* concluded that the Army "does not need or want our collaboration as amateurs." "No, our field is not the military," wrote the spurned editor, "We are civilians. Our ARRL is a civilian organization." With the majority of hams denied genuine military roles, the League resigned itself to naming civilian defense "our primary field in the defense picture."[16] Occupation of this middle ground served hobby radio well as a public relations strategy into the Cold War.

Cold War Suspicions

The FCC allowed hams back on the air after World War II ended, though radio continued to attract media attention as a risky technology. When hobbyists interfered with Army communications during a Korean War battle, the *New York Times* ran the story on the front page: "Radio Hams in U.S. Discuss Girls, So Shelling of Seoul Is Held Up." The reporter emphasized that the conversation between a hobbyist in Seattle and another in Portland was particularly frivolous—including mention of a "date for the movies in the evening and some basketball scores"—compared to the transmission it interrupted, a colonel giving orders to a tank commander.[17] The incident was dramatic, but a fluke that resulted in only a brief disruption.

The potential use of two-way radio in espionage caused much greater concern. Wireless technology seemed ideal for clandestine communications—mobile, long-range, and relatively easy to operate. The only drawback was that anyone could listen, and the FCC always had banned hobbyists from sending "secret codes or ciphers with hidden messages" in the hope that the lack of privacy would discourage illicit activity. In the late 1940s and into the 1960s, anti-Communist hysteria fueled suspicions about hobby radio. Friendships with foreigners, according to a prevalent anxious logic, might lead hams to yield to an authority outside the United States and perhaps even commit espionage. One woman worried in 1956 that her whole family had "become suspect and is shunned by polite society" because of her husband's international hobby. Cold War films and fiction depicted criminals using radio, and gadgets for covert communication appeared prominently in the more than thirty television series of the 1950s and 1960s that featured espionage. Whether inspired by geopolitical fears or glamorized fantasies, the theme of spying seeped into the public imagination to the point that in the mid 1960s neighbors supposedly turned to hams, known as local electronics experts, for help detecting concealed surveillance devices.[18]

Hobbyists inadvertently may have contributed to associating radio with subversive activity by recounting heroic stories of patrolling the airwaves. Decades after the Secret Service had sought surveillance help from a New Jersey amateur radio operator who owned a high quality receiver, hams continued to promote the triumph of home-brewed technology. In 1915 the hobbyist had made phonograph recordings of exchanges between "a German subsidized radio station" on Long Island and an operator in Germany, and Secret Service agents found that the messages contained encrypted information about the movement of supply ships bound for the Allies. Once the spy on Long Island was in federal custody, "the sinking of ships by U-boats fell off sharply." The founding documents of the Radio

League of America written later that year made surveillance sound like a routine hobby experience, stating that one purpose of the club was "to check on and report such activities of German agents in this country as they might hear." In 1941, *This Week* magazine described hams "patrolling the ether day and night." The access of hobbyists to the airwaves purportedly guaranteed that "any attempted espionage work by radio in this country will enjoy a startlingly brief existence."[19] Of course, a nervous Cold War public realized, this access just as easily could be put to other ends.

International intrigue surrounded postwar ham radio as portrayed in the press. A 1949 article in *Time* magazine mentioned that "Every week, U.S. hams casually talk to hams behind the Iron Curtain." Although acknowledging that "Usually the topics discussed are politically innocuous," the story hinted at tantalizing information conveyed in such conversations. The postcards exchanged to confirm ham contacts aroused further suspicion. *Time* reported that "government-made" cards sent by "Red hams" displayed "propagandistic puffs for Russian greats" and quoted a Romanian who had added a "chatty note" saying his neighbor "had just been arrested." Hobby magazines, too, occasionally exhibited Cold War paranoia. *CQ* defended the increased FCC oversight as protection against "espionage." In the words of a ham questioned by the FCC for communicating with an illegal operator, anonymity on the airwaves meant "You never know what trouble your friends can get you into."[20]

Confirmation cards with photographs of distant lands, greetings in uncommon languages, and exotic stamps contributed to the excitement of long-distance ham radio communication and piqued the curiosity of nonhobbyists. A ham's wife reported feeling that "The mail-man eyes me suspiciously as he hands me colorful post-card things scrawled with a queer jargon." Ham radio organizations established mailing centers to consolidate the sending of confirmation cards. Hams who used such clearing houses then received a sealed envelope enclosing several postcards. Primarily amateurs explained mail consolidation as a way to cut postage costs, but they also noted that this system reduced scrutiny of mail exchanged between radio hobbyists in the United States and in Iron Curtain countries.[21]

News reports on spy cases played up any connection to radio communications. By coincidence, the deportation from Mexico and subsequent arrest of Morton Sobell by the FBI in August 1950 occurred the day after the arrest of Spanish refugee Enrique Ricart Corts by police in Mexico for transmitting information to Russia by radio. Though multiple wire stories hinted of a possible link between the Sobell and Corts cases, this proved unfounded. The media described Sobell as an engineer and "naval radar expert" wanted in connection with a "Russian spy ring," for which he later faced trial alongside Ethel and Julius Rosenberg. In his autobiography, Sobell acknowledged the broad association of electronic devices with espionage. The Mexican police who packed the Sobell family's possessions allegedly stole several items. As the FBI examined the contents of his luggage, Sobell noted missing "the small piece of electrical equipment, which I had taken with me as a sort of conversation piece in case I met any electrical engineers." "I had kinder thoughts about the petty-thieving Mexican police when I realized the FBI had not found that synchro in our luggage!" recalled Sobell, for he was "sure it would have been a star exhibit at the trial."[22]

Scrutiny of radio hobbyists intensified as the anti-Communist movement gained prominence through the actions of Senator Joseph McCarthy and the House Un-American Activities Committee. In November 1953, Senator Alexander Wiley, Chairman of the Foreign Relations Committee, called for tighter regulation of hams for security reasons. He specifically cited the possibility "for a disloyal operator to guide a Soviet plane to its target" and assist in an attack on the United States. When the editor of *CQ* responded that hams possessed an "overwhelming national loyalty" and "act as their own policing system," Wiley conceded the good behavior of many in the hobby. Still, he said, this did not change "the fact that the Communists are keenly aware of the significance of amateur radio for their treacherous operations." The Subversive Activities Control Board already had presented evidence to Congress of the Communist Party's attempt to establish radio communications, which had included "a search [undertaken by the Party] to find amateur radio operators among CPUSA members." Joseph McCarthy declared that "the Hams are a tremendous potential for passing out improper information for espionage and so forth" and went on record in support of a bill "to require TV and Radio stations, including amateurs, to record all programs and transmissions." Caught up in the anti-Communist furor, the FCC proposed a new restriction in June 1954 that would have made "ineligible for licensing any amateur or commercial operator who is a member of the Communist Party or any organization which has been required to register as a Communist-action or Communist-front organization under the provisions of the Internal Security Act of 1950." Application forms were to include questions about party affiliations, and each prospective licensee would need to file a set of fingerprints with the FCC. The FCC debated the proposal into 1956, but the political mood changed before it could be enacted.[23]

Direct international contacts exposed hams to information unavailable to most Americans. Where non-hobbyists mainly saw risks in open communication with foreigners, radio enthusiasts saw the chance to gain unique insight into politics abroad and perspective on the Cold War rhetoric surrounding them at home. One hobbyist recalled that, hearing reports broadcast by Radio Moscow in the late 1950s, he "wondered how those Russians could be as bad as our U.S. propaganda of the time said they were." If a radio hobbyist wanted to know "what sort of guy" the average Soviet ham was or "Did he have to be a Party member to get a license?" he could simply ask a Soviet over the air. Based on his own conversations with Soviets, an American hobbyist believed ham radio could serve "as a means of improving relations between the United States and the Soviet Union." Hams who focused on long-distance contacts, argued a 1964 editorial in one club's newsletter, contributed to "the furthering of world cooperation through communication, and I mean down to earth communication with our fellow man."[24]

Praise of unrestricted communication combined with the political neutrality prescribed by the technical culture of ham radio sometimes made hams appear far removed from mainstream American culture. Fred Huntley drew vocal criticism when he organized the Anti-Communist Amateur Radio Network in 1961. His fellow hobbyists found Huntley's goal—to use the "large untapped potential [of ham radio] for alerting the nation on the dangers of communism"—to be a violation of the community's apolitical stance. "We are doing a much better job of corrupting Communism by just being ourselves and talking with the Russian hams than we could ever do with an overt attack through ham radio," responded the editor of *73 Magazine* in a statement that won the support of many hobbyists.[25] But any group opposed to anti-Communist action during the Cold War had some explaining to do. Instead of trying to convince inward-looking

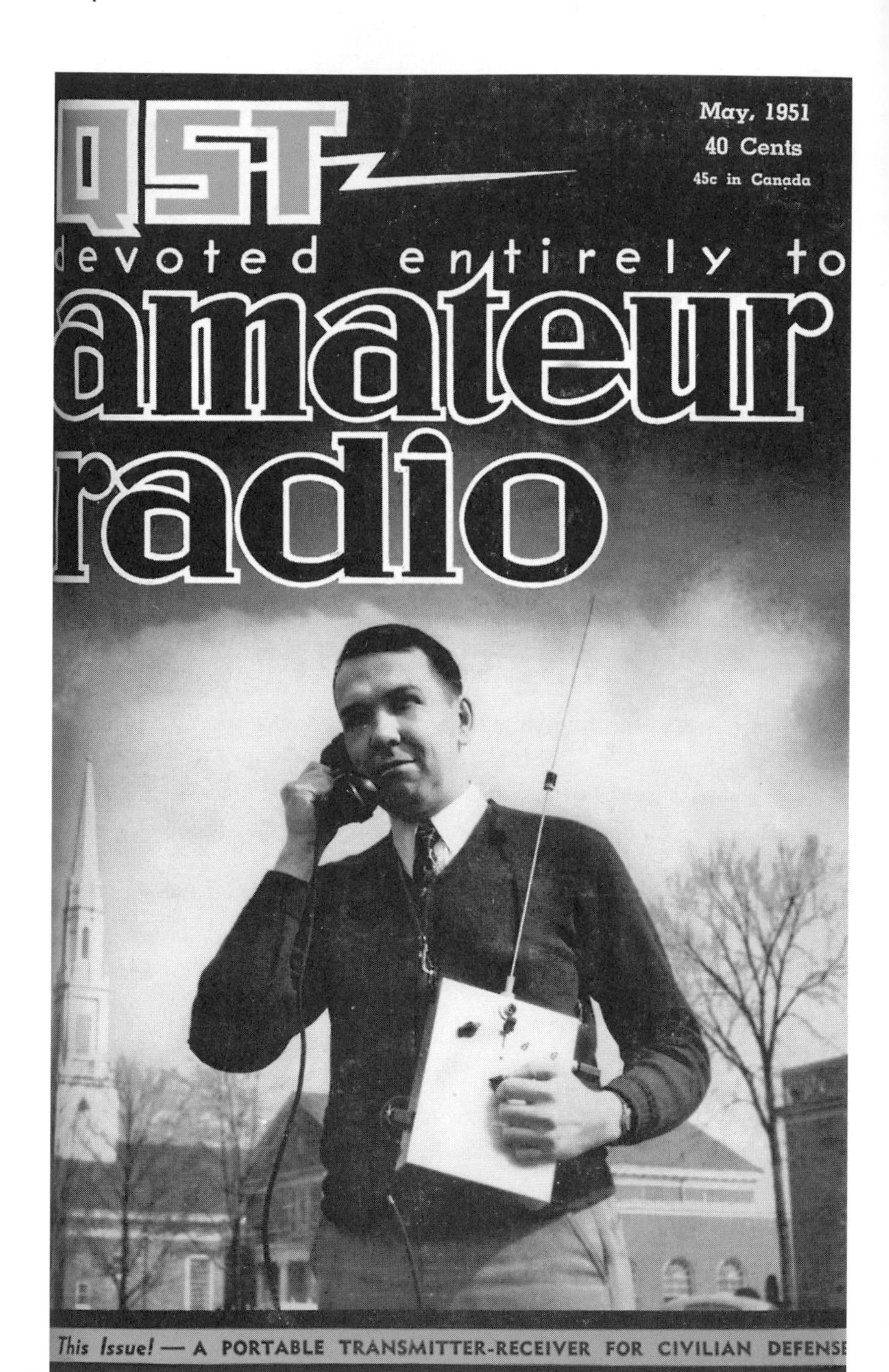
QST
May, 1951
40 Cents
45c in Canada
devoted entirely to
amateur
radio
This Issue! — A PORTABLE TRANSMITTER-RECEIVER FOR CIVILIAN DEFENSE

Americans that communicating with foreigners might help smooth over political differences, hams shifted attention to activities considered more appropriate.

Public Service and Public Relations

Hobbyists blamed insufficient public relations work before the United States' entry into World War II for the four frustrating years of silence they endured, and they took from this the lesson that publicity must occur continually and must anticipate any challenge to the hobby. Businesses dependent on hams joined in the cause. Why create a new hobby magazine in 1945, with the transmitting ban still in effect? Publicity. "By starting now," editors explained in the first issue of *CQ*, "we shall be in the most advantageous position to cooperate with every individual and organization in securing adequate postwar recognition of the amateur and his requirements." General Electric, a leading supplier of components to hams, offered the Edison Radio Amateur Award annually beginning in 1952 to recognize "meritorious public service." This pleased GE's customers—ham magazines proudly reported the results—and associated the manufacturer with positive aspects of the hobby.[26] Public relations became so much a part of ham radio that it was fundamental to the hobby culture.

When ham radio fell under close scrutiny during the Cold War, hobbyists publicized their role in civil defense programs to divert attention from conversations perceived as possibly subversive. The establishment of a secondary communications system to be used in the event of an attack on the United States grew out of the same Cold War anxieties that made neighbors question the intentions of hams who chatted with foreigners. Volunteering their technology to assist the Federal Civil Defense Administration placed hobbyists in quasi-military roles. (With the shift from war to peace, the government changed the terminology for its militaristic citizen programs

Figure 5.2
During the Cold War, hams' conversations with foreigners provoked suspicion. Service as civil defense radio operators—prepared to handle emergency communications following an attack—focused attention on a benefit of hobbyists controlling powerful radio technology. *QST*, May 1951 cover, reprinted with the permission of the American Radio Relay League.

from "civilian defense" to "civil defense.") The virtues of ham radio for community protection then counterbalanced its risks. In May 1951, a photograph on the cover of *QST* showed a serious-looking ham operating a portable radio on an idyllic New England town green, complete with a white church steeple in the background (figure 5.2). An article inside gave detailed instructions for building the device, designated for civil defense, and claimed it would help hobbyists meet "a new need born of the atomic age."[27] This was just one instance among many in which the American Radio Relay League played up the rather limited involvement of hams in civil defense to portray recreational radio as far more than a hobby.

The ARRL had set about crafting a role for ham radio in national security after World War II to prevent any future curtailment in operating privileges. When the Department of Defense created the Office of Civil Defense Planning (OCDP) in 1948, an editorial in *QST* assured readers the League was "already in touch" with the new civil defense authority, "with a view to knitting our activities into the national needs." The OCDP was exploratory, established to investigate the need for a permanent agency that would study how to respond in the event of an atomic strike. The findings presented in the Hopley Report (its name drawn from OCDP director Russell Hopley) remained influential for the next decade until pacifists' protests raised public doubt about the call to civil defense. To guarantee that "in any emergency, communications in some form will be available," the OCDP described the need to prepare for "every contingency." The Hopley Report recommended layering technological systems to achieve reliability even under attack.[28] This planned redundancy opened the door to ham radio.

The ARRL had managed to place two of its executives on the communications advisory panel to the Office of Civil Defense Planning, and the voice of the ARRL came through clearly in the Hopley Report. "Emergency service is a tradition in amateur radio operations," began the paragraph on ham radio in the report's communications section. "Under a carefully organized plan they [licensed hobbyists] are capable of making an important contribution to civil defense in providing supplementary emergency communications channels, especially during a post-raid period." Despite this apparent endorsement of radio operators' potential for public service, the Hopley Report outlined a peripheral role for hams in civil defense com-

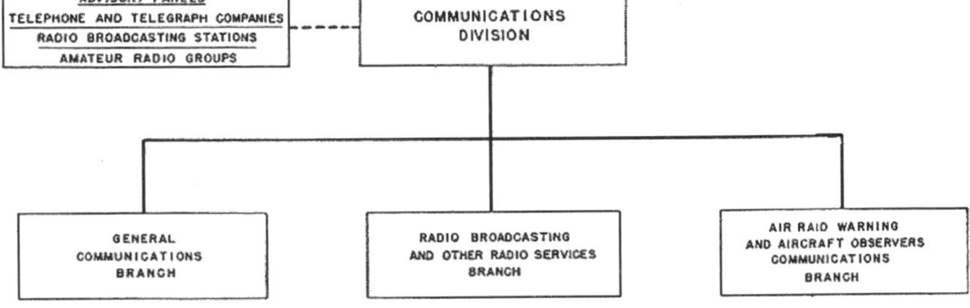

Figure 5.3
The plan for civil defense communications granted ham radio groups an advisory status, but indicated their peripheral relationship by connecting them to the central division with only a dotted line. Chart 8 from U. S. Office of Civil Defense Planning, *Civil Defense for National Security* (the Hopley Report) (1948).

munications. The proposal classified amateur radio groups along with communications businesses as mere "advisory panels." In the chart showing the structure of the Communications Division, the advisory panels were held off to one side and connected only by a dotted line, while solid lines joined the principal branches (figure 5.3).[29]

Once the creation of the Federal Civil Defense Administration (FCDA) in 1950 confirmed civil defense as a national priority, the ARRL described ham radio's emergency communications as part of an unspoken bargain for the right to operate. According to a *QST* editorial on the matter, "The question—the big question—is not whether we are able to furnish radio communication but whether we will be permitted to do so." Fearing that hams might spend precious time, money, and energy on preparedness only to be shut down again in the event of an actual crisis, the League demanded the FCDA answer this question before recommending that hams become involved in civil defense. In the meantime the ARRL told hobbyists to think about innovative methods to reduce the security risks of wireless communication. Such a demonstration of technical proficiency, the League figured, would offer "the very best insurance of our being fitted into the permanent civil defense picture."[30]

Despite the efforts of recreational radio organizations, hams were only ever granted a marginal role in the official civil defense plan. The FCDA deemed radio too insecure for a primary communications system. Civil defense policy analysts repeatedly pointed out that the dangers of message interception and the detection of transmitter location made wireless communications "particularly vulnerable." In late 1951, the FCDA appeased hobbyists by establishing the Radio Amateur Civil Emergency Service (RACES). The Hopley Report had stressed that secondary systems should be the responsibility of state and local civil defense organizations.[31] By providing them with local backup duties, RACES satisfied hams' desire to participate in civil defense and minimized the threats of their wireless exchanges to the federal system. The relegation of hobbyists to second-tier communications demonstrated the FCDA's faith in the reliability of radio machinery—hams' pet technology was expected to outlast the ubiquitous telephone—yet it also undercut their power.

The extremely limited mission of RACES kept hobbyists out of the way while professionals undertook the tasks central to civil defense. RACES activities primarily consisted of enrolling members—after checking their "loyalty to the United States and general reliability" as evidenced in "police, employment, and scholastic records"—and performing one annual emergency communications test. The FCDA acknowledged that "The radio amateurs provide a valuable source of skilled communicators and emergency communications equipment for civil defense," but RACES existed for redundancy. When the program was ten years old, its governing agency clarified that RACES was "intended only to supplement any established local communications systems, not to replace them."[32]

This was hardly the role that hams had dreamed of. The ARRL's language of "knitting our activities into the national needs" and "our being fitted into the permanent civil defense picture" had projected a false modesty, a willingness to be wholly subsumed. Though hobbyists did not invent the role of militaristic citizen, but rather just introduced a technical component to the existing civil defender role, they did expect the modified radio operator–civil defender identity to bring special attention to hams. The ARRL resented that RACES instead stripped away hobbyists' distinct identities. Operating in a local RACES unit, the League complained, "Each participating amateur would be required to subjugate his amateur identity—a disadvantage from the morale standpoint." Hams' dissatisfaction with the arrangement led to low enrollment, which further debilitated the program. Roughly 1,400 communities had filed RACES plans by the start of 1961, far short of the 5,000 groups that the Office of Civil and Defense Mobilization believed necessary for a thorough backup system.[33]

Quietly, the ARRL agreed with the FCDA that RACES served little purpose. The League called the program "almost indistinguishable from regular amateur network operations" and blamed "the tardy development of national planning and regulations" for the fact that so "much steam has been lost from amateur interest in civil defense work." A three-part series in *QST* on the status of RACES in 1953 raised a number of criticisms of the service. More than a decade after the creation of WERS, the ARRL still held a grudge that WERS had not been primarily "an *amateur* service." The League was frustrated that RACES similarly kept hams from controlling their own service activities. "We amateurs, *as* amateurs, can do nothing to organize civil defense," the ARRL humbly conceded with regard to RACES. "We can only provide a radio communications service for a civil defense organization." In its final analysis, the ARRL capitulated to the FCDA for the good of the hobby. "To have amateurs and government lose cordial contact with each other on the matter of RACES," the League realized, "would be disastrous, and we do not intend letting this happen."[34] The ARRL retained hope that having a system for emergency communications in place might earn hams respect and protect their operations in future crises.

Publicly, the ARRL supported RACES as an extension of hobbyists' broader volunteer work in civil defense. Frequent *QST* articles on civil defense activities before RACES existed had portrayed hams as in control of emergency communications. Headlines recounted how the "Motor City Amateurs [Formed] a Vital Link in CD Communications" and the "Tri-County Radio Association Program Provides Emergency Stations and Promotes V.H.F. Activity in Northern New Jersey." The League encouraged hobbyists to follow in the footsteps of these exemplary clubs and published instructions for building equipment specialized for the civil defense frequencies. As always, the ARRL linked service and public relations. In a farce of a report by a ham club's inactive civil defense committee, chairman "O. Y. Bother" admitted that "since we never did bother about those publicity releases the League sent us" the local civil defense communications director "had never heard of us." For the good of the hobby, the ARRL advised hams to view the federal expectations presented by the "challenging development" of RACES as a test. "We must show that we can do a responsible job when given the chance," chided a 1952 editorial in *QST*. "*How* we embrace opportunity *individually* will determine the standing of our service in the future."[35]

The FCDA's plan for national security counted on the preparedness of individual citizens and especially family units. *Life* magazine's insistence that "there is much that you can do to protect yourself—and in doing so strengthen your nation" invoked typical civil defense rhetoric. Following from the description of civil defense as a personal, moral obligation, hams' participation enhanced their image as individuals in addition to improving the perception of the hobby overall. A handbook published by the ARRL called attention to this benefit, describing the annual Simulated Emergency Test as "a subject for good local publicity for amateur radio, and for all those who take part." As interest in civil defense peaked, *Life* relabeled "The man down the street with a backyard shelter [who formerly] was considered odd" as "actually a solid, sensible man—and a responsible citizen."[36] Civil defense service similarly transformed hams from outsiders with a fondness for strange contraptions to patriots in command of survival gear.

Although hobbyists played a minor role in the national civil defense plan, their strategy of promoting the service in exchange for airwave rights succeeded. Individuals who volunteered as emergency communicators earned special rewards. When the scarcity of electronics components during the Korean War restricted ham building projects, those who belonged to RACES or any of eight other recognized military or civil defense organizations received a doubled annual quota for supplies to maintain their stations.[37] The hobby community overall won the privilege to continue open, recreational two-way radio communication throughout the Cold War.

After achieving a strong position for Cold War ham radio, the League relaxed its commitment to federal civil defense programs. The number of states reporting hams' participation in Operation Alert, the FCDA's nationwide readiness test, held steady between twenty-five and twenty-eight in the second half of the 1950s. Then in 1961, hobbyists in only ten states participated.[38] Meanwhile the popularity of the League's own readiness test grew. The ARRL distinguished its annual Simulated Emergency Test as belonging to hams, saying "it's not part of a big government-sponsored project, like Operation Alert. It's our own activity, using our own amateur-sponsored and amateur-led organization." When involvement in the 1961 ARRL Simulated Emergency Test "exceeded [the] 'RACES boom high' of 1952 for the first time," the League declared that "Dependence on the government for direction is gradually giving way to dependence on *ourselves*."[39] The Simulated Emergency Test granted hobbyists autonomy and afforded another important advantage by concentrating hams' public service on general emergency communication.

The federal civil defense plan received only limited public support and inspired very little participation. Local civil defense branches struggled to rouse unresponsive communities who did not feel threatened. Even during World War II, the relative safety of Americans living free from enemy attack had led to a decline in cooperation with preparedness drills by mid 1943.[40] Mere talk of a threat generated less fear during the Cold War than during wartime. Despite admonitions from the FCDA throughout the 1950s, the strongest show of interest in civil defense came not until the Berlin crisis and Cuban Missile Crisis of the early 1960s, and even then it was scattered and short lived.[41]

Citizens' indifference toward civil defense meant that the ARRL's public relations campaign pitched to federal regulators largely failed to ease the conflicts hams had with their neighbors about electrical interference and the appearance of antenna towers. To better represent ham radio in their immediate communities, individual hobbyists and clubs adapted publicity strategies suggested by the League. Instead of focusing on civil defense, clubs emphasized emergency communication more broadly. This generalized service had greater practical appeal than did civil defense, making it an appropriate choice for hobbyists seeking local support.

Hobbyists always had been eager to provide communications when other systems failed. This arrangement proved mutually satisfying: communities left without telephone service after a natural disaster received vital messages, while hams demonstrated the reliability of radio technology and of radio hobbyists. Before the institution of civil defense, many community ham clubs already had standing committees that organized emergency response drills. Most hobbyists viewed the role that the ARRL proposed for radio in civil defense as just a single facet of emergency communications, with the main difference being that the kind of disaster causing the disruption in normal service, nuclear attack, was specified from the outset. The Rochester Amateur Radio Association was one of the clubs that simply expanded its preexisting emergency response group to include civil defense responsibilities.[42]

Hobbyists approached general emergency communication conscious of its value for public relations. A 1935 article on "How to Gain the Goodwill of the Public," for example, advised that the voice of a hobbyist coming over a neighbor's broadcast radio receiver would be less likely to ignite an argument if the listener "recognized it as that of one of the boys who had rendered him a service," and the hobby press continued this argument into the 1960s.[43] Explicitly describing it as a way to gain outsiders' support, *CQ* began offering an annual award for the club "which makes the greatest contribution to the community in an emergency." The ARRL told clubs performing emergency drills that inviting the press was "one of the best ways of exciting public interest and creating an awareness that amateur radio is really doing something of public benefit."[44] For its 1961 Simulated Emergency Test, the League admitted a dual objective, "first, to test our emergency potential and capability, and second, to give a public demonstration of our abilities." When "the vast expansion of commercial communications systems" appeared to undermine the plausibility of hobby radio as a form of emergency communication in the mid 1960s, the editor of *CQ* wondered, "How then do we convince our public that we are a necessary and vital part of community life?" and immediately followed, "We must do it through a carefully planned public relations program." His answer focused not on alternative ways to assist after a disaster, but rather on a new publicity tactic.[45]

Hams with civil defense insignia on their equipment, of course, remained capable of providing backup communication during any disruption of regular systems. Since the press began trumpeting hams' potential service in national security in the 1920s, the only examples of actual emergency communications by civilian hobbyists had come following natural disasters. No unit of RACES ever went into action for defense purposes, yet the RACES station in Anchorage, Alaska, for instance, stood in for regular communications channels wiped out by an earthquake in 1964.[46] While the ARRL and other ham radio organizations continuously promoted emergency communication as the hobby's community service, local clubs routinely downplayed civil defense. Even the ARRL stopped associating emergency communication with civil defense in the 1960s as Americans' support for civil defense programs plummeted. A handbook for radio hobbyists then reported that though "the over-all Civil Defense program of the country has been dragging its feet, because of public apathy, the radio facilities furnished by the ham fraternity are quite efficient and prove their value every time a public emergency arises."[47]

The ham community addressed national and local concerns through a split public relations strategy. The ARRL focused on staying on good terms with the FCC and on lobbying Congress for legislation protective of amateur radio. From the League's perspective, hobbyists should volunteer for military service and civilian communication programs as a way of fulfilling commitments the ARRL made in bargains with the state. On an individual and day-to-day basis, hams needed to negotiate with neighbors and municipal governments. That local constituency more readily appreciated hobbyists' emergency communications programs for responding to natural disasters, which were untainted by the political rhetoric of national civil defense.[48]

Hobbyists seeking favorable publicity appealed to different aspects of the same Cold War culture that raised suspicions about ham radio. A pastime based on open, international communication violated the cautious climate of political and social containment. Hams who advocated improving relations with the Russians through individual contacts were a tiny minority that stood no chance of overturning the militaristic stance taken by government officials and propagated through the news media and popular culture. The technical culture of the Cold War, however, provided an entry for radio hobbyists. In preparation for a conflict anticipated at some unknown future time, the military-industrial complex prized reliability and redundancy in devices and systems. Radio was not the first choice for communication, but second-place technologies carried weight in mid century America. Taking on a role in backup communications made hams seem dependable and strong and made radio a more acceptable Cold War hobby.

6 Ham Radio at Home

Hams domesticated radio technology just enough to find a place in the home. They toned down radio's militaristic image and declared ham radio an amateur pursuit rather than an extension of work. To keep the technology special, hobbyists avoided the kind of complete standardization of its appearance and function that would have made two-way radio accessible to all. The identity radio hobbyists shaped for themselves through technology similarly exhibited a limited domestication. As members of a technical fraternity, hams stood apart from the household. They continually referenced the family, however, and tied the hobby to the home through remarks about domestic conflicts brought on by radio.

Hobbyists gained independence in the home as a result of the disruptions caused by their use of a partially domesticated technology. Rough, bulky two-way radios, prone to interfere electrically with television and radio receivers, were best kept isolated. In staking out hobby spaces, hams gained privacy as well. A 1941 *Harper's Magazine* article that focused on "the amateur's services to society" concluded with the assertion that "it would be shortsighted to ignore the personal cultural value of amateur radio." The authors explained that "Amateur radio gives to ordinary men, leading the circumscribed lives of ninety out of a hundred people, a release from humdrum existence and routine compulsions; it makes them freer men."[1] This was a socially sanctioned escape, the temporary relaxation provided by a hobby safely contained within a domestic context.

Social Disruptions

In the mid century home, the implications of ham radio varied with the age of the participant. Families demanded few responsibilities of boys and approved of a wide range of youthful pastimes. It made little difference to family life whether boys spent their free time playing baseball or doing ham radio. Before the age of electronics, boys involved with two-way radio often had been labeled pranksters. But even then the "small boy given to tinkering with radio in such a manner as to [...] superimpose a staccato of clicks and buzzes on the pianissimo passages of Beethoven's Fifth" was no more maligned than the boy whose stray baseball broke a neighbor's window.[2] With the increased value placed on technical skills following World War II, educators told parents to encourage sons' technical hobbies, which might be career stepping-stones. At mid century the public perception of electronics as innovative lent cachet to the tinkering of boys and young men. The Cold War urgency to produce scientists and engineers further imparted a sense that boys active in ham radio had the potential to strengthen national security.

Complaints voiced by families about the amateur radio activities of boys appeared very infrequently in the hobby literature compared with complaints directed at adult ham operators. The single article published in the major ham magazines in which "A Ham's Mother Has Her Say" stood out against dozens of analogous articles that offered a "Wife's Eye View" and bemoaned the "Ham Shackles" of marriage. Boys ignored chores, missed meals, and kept messy bedrooms because of radio activities. And parents of hams grew frustrated with hobby jargon they did not understand and worried that their sons risked electrical injury.[3] Yet these irritations and concerns typically were dismissed as a natural part of play.

Between 1947 and 1962, four articles in *Parents Magazine* advocated tinkering by boys. Each carried the message that industrious hobbies could prevent children from becoming idle and mischievous. In the first, Joy Freed recounted the turning point in her attitude to her son's hobby of building model airplanes. Seeing some teenage boys misbehaving had made Freed wonder "why they were on the street at that time of night, and if their mothers didn't care." Suddenly her son's untidy room seemed insignificant, and Freed dedicated herself to supporting his hobby. She "resolved" to "crawl on my knees if necessary picking up pins and sticks" for the sake of "keeping my boy happy and busy at something worth while." Another of these articles promoted ham radio in particular as having "the enthusiastic endorsement of PTA, churches, social service groups and law-enforcement agencies." Parents were told to tolerate their sons' ham radio activities because they functioned "as a deterrent to juvenile delinquency" and might lead to successful careers in the sciences.[4]

The pieces in *Parents Magazine* blamed parents who interrupted tinkering boys, or attempted to confine their projects, for interfering with their sons' educational development. As an example of how "overly tidy mothers or noise hating fathers" caused "many children [to] lose interest in science," a 1962 article told of a boy who failed to build a radio "because his room was thoroughly 'straightened-up' every week." This was contrasted to the experience of a mother who cheerfully abided the mess in her son's room, including a "disemboweled radio," and then proudly witnessed his winning a science fair. A 1955 story characterized meddling mothers as an obstacle to learning over the airwaves, recounting fifth grader Bobby Fiske's correction of a smart female classmate on a geography fact with information learned by talking to someone in Liberia the day before. Bobby claimed that he "would have found out a lot more, only then Mom called me to supper." These critiques of parents who intruded upon technical hobbies echoed the broader accusations heard at mid century that overbearing mothers emasculated their sons, an idea Philip Wylie termed "Momism" in his 1942 book, *Generation of Vipers*.[5]

Attempts to persuade parents that sons should be allowed to tinker highlighted the possibility that boys engaged in technical hobbies later might attain respectable, secure positions as scientists and engineers. Technical hobbies were reported to contribute to the development of "such desirable character traits as persistence and ingenuity." A more direct connection came in the suggestion that in addition to "enjoying himself and learning about chemistry, too!" a boy shown using a chemistry set might "someday" become "a professional chemist." The wordplay in the title of an article published during the space race, "A Space Program for Your Young Scientist," offered parents the tantalizing hope that making room for technical hobbies might help them rear astronauts.[6]

The pressure on parents to embrace technical hobbies extended beyond considerations of sons' edification and future careers to the pressing matter of national security. In 1955 *Parenting Magazine* admonished Americans that "In an atomic age, our national survival may well depend upon having a reservoir of trained technicians in peace and war" and portrayed youngsters involved with electronics as initiating vital training. *Parenting Magazine* was just one voice among many calling for the support of boys in scientific endeavors. When Industrial America of Chicago released five technical hobby kits, for example, the Under Secretary of Commerce praised the kits for providing "genuine stimulus" to technological advancement and economic expansion, and thus contributing to the "preservation of our freedoms." *Fortune* magazine concluded from these remarks that "Responsible parents, whatever their misgivings, obviously will have no choice but to see that Junior gets the whole works." General acceptance of boys' radio hobbies continued throughout the 1970s, tied to the prospect of ham radio leading to scientific and technical employment and helping participants avoid "waywardness."[7] These benefits also eased acceptance of all hobbyists' activities. But the homosociality of ham radio, which fit the pattern of boys' leisure, proved problematic for grown men.

To the extent that it grounded masculinity in technology rather than in sexuality, ham radio threatened social relationships. Many adult hobbyists described radio activities as incompatible with romantic interests. According to one source, interest in ham radio followed "a fairly uniform pattern" of varying inversely with sexual desire. "The youth, at first completely absorbed, gives up radio when the opposite sex begins to compete seriously with the fascination of microphone and key. This alienation normally lasts until the first baby is born, at which time recrudescence sets in and he is a ham all over again." Recounting his own turn from radios to women, a ham in 1950 called it "the old, old story—the hobby which wore skirts won out." "I married this little hobby and with her now tucked safely under my arm I felt free to go back to my original love," he said of his return to radio.[8] Men who appeared more interested in radio equipment and conversations with other men than they were with the opposite sex faced harsh rebuke and thinly veiled questions of their sexuality.

In the immediate postwar years and through the 1950s, conflicting opinions regarding the obligations of men to family and to self bombarded the

American public. Mass media outlets ranging from fictional television programs to feature articles in magazines celebrated the middle class nuclear family composed of married parents with children. These depictions of husbands and fathers linked the power to lead the family unit to a responsibility to support it financially. Men were instructed to be good "breadwinners," devoted providers to their families.[9] At the same time, psychologists diagnosed a crisis of identity among men. Preoccupation with work and family responsibilities, the experts warned, had caused men to lose their sense of self. Loss of identity seemed to endanger free will and consequently raised alarms at a time when political ideology staunchly opposed collective thought.[10] Psychologists' advice that men should protect their identity as individuals directly contradicted popular culture's message that men should make family life top priority.

Reports of disagreements between male ham radio operators and their families indicate how these competing public pressures played out in private. Complaints that radio upset home life filled the hobby literature and became a standard trope of hobby culture. General Electric acknowledged that active hams let family duties fall by the wayside when it presented the wife of the 1953 Edison Radio Amateur Award–winner with a gold watch for being "the most understanding wife of the year."[11] How accurately the motif of spousal bickering reflected actual experiences is difficult to determine, since it seems to have served partly as a reminder to outsiders that hams were married. Regardless, these stories of disputes about ham radio did encapsulate the debate over whether a man's first responsibility should be to self or to family.

Domestic arguments about ham radio erupted around the allocation of the most basic resources, time and money. Operating a recreational radio station imposed a considerable financial burden on the family while benefiting only one individual. Equipment prices and many little extras like magazine subscriptions, contest fees, and club dues amounted to quite a sum, even within the budget of the middle class ham. A 1957 survey found that the average reader of *CQ* magazine earned $7,350, valued his present equipment at just over $1,000, and expected to spend $245 (one thirtieth of his salary) on the hobby in the coming year. The price of a single piece of equipment could be daunting. The Johnson Ranger, a mid level transmitter, cost $293 in 1956 and $360 in 1965. More powerful models and those from more prestigious manufactures cost far more. The Collins KWS-1 transmitter sold for just under $2,000 in 1956. And a transmitter was only one of several pieces of hardware essential to the home station. A hobby receiver in 1960 could cost as much as "a 21-inch television console, a new furnace for your house, or a fairly decent used car," despite the fact that, a handbook for new hams pointed out, the apparatus normally was "rather severe looking" and came without a speaker. The radio literature told of married hams enmeshed in "spats about spending $9.98 for a special condenser" instead of allocating the money for household items such as "grass seed or the grocery bill." Noting that hobbyists occasionally avoided debates about the family budget by sneaking newly purchased equipment into the house, R. W. Johnson explained, in verse, how this type of dishonesty soothed tensions: "For his wife was not aware of where the money went, All she knew was quite untrue so she was happy and content."[12]

According to the conventions idealized by the media, all husbands owed some after-work hours to chores around the house and the care of children. Household work of middle class men in the 1950s included do-it-yourself maintenance and improvement projects, which were thought of as a kind of hobby. In deciding when to stop the endorsed hobby of chores and begin freer leisure, men set the limits on their obligations to others and to themselves. A typical complaint voiced by one radio hobbyist's wife was that "the darn 'Set' was such a nuisance when mealtime came or when she needed a few little things done around the house." Another ham alluded to the tension inherent in such decisions when he invoked the rhetoric of clashing hobbyists and wives to explain his inability to edit the club newsletter on time, saying "the XYL [wife] handed me a list of 'things to do' around the yard and house this summer that I just can't put off any longer." A few years earlier he had blamed "the confusion resulting from getting the kids packed for the trip to Grandma's" for the fact that the summer issues were "taking a beating!"[13]

Hams spoke of hobby participation as out of step with fatherhood. The hobby slang for children, "harmonics," played on a radio term for multiple frequencies born of a fundamental frequency, which had the potential to disturb clean transmission. A member profile in a club newsletter mentioned he had "one YL [female] harmonic not quite a year old but already interfering with daddy's DX [long-distance operating] activities." Hobby publications repeatedly listed "screaming children" as a motivation for seeking sanctuary in the home radio station. According to a 1960 handbook, a typical ham's evening followed the pattern of, "You come home from work, eat your dinner, maybe glance at the headlines, then make a beeline for your shack." Hobbyists who sought time for radio and for parenting sharply divided their schedules. One common split was to confine radio activities to the hours when children slept. Some hams made a seasonal trade off, like the hobbyist described in a club bulletin as someone who was "vy [very] much the family man but has got a vy fb [excellent] tower es [and] beam in the backyard for winter." Grievances about the difficulty of reconciling ham radio with family life ranged from callous complaints—a ham who griped that his son's two week hospitalization caused him to miss the chance to communicate with a rare station from Brazil—to mild grumbling—a hobbyist who told his local radio club that trying to compete in contests on two consecutive weekends was "a little tough" and wondered whether the organizer intended the events "just for young and single sprouts."[14] Whatever the tone, hams' discourse counterposed hobby life and home life.

Amateur radio did more than tempt men to squander household money, ignore chores, and spend insufficient time with children. In language that operated as the gendered parallel to men's hobby rhetoric, women wrote of resenting ham radio for weakening emotional and physical marital bonds. Radio magazines in the 1940s and 1950s regularly published strongly worded commentaries by hams' wives, including protests that technical interests diminished hobbyists' sexual interests. There was an element of satire in these essays, but the consistent pattern of remarks suggests this was joking about a genuine, if exaggerated, concern. When men chose to talk via radio to other hams instead of in person to their own wives, women reported feeling competition for their husbands' attention. Several articles protesting the hobby's interference with intimacy appeared in radio magazines in the decade after World War II. The frankness with which women wrote of their desires can be attributed partly to the reigning social expectations of wives. Conveying early Cold War sexual-political anxieties that associated homosexuality with Communism, psychologists and popular culture alike posited married women's sexuality as essential to the heterosexuality of their husbands and sons and, following from this, to the political stability of the country.[15]

Television, the electrical technology perceived to be at the center of family life, also drew criticism for disrupting marital sexual relationships in the late 1940s and 1950s. Whether the image that held a viewer's attention was a beautiful woman or a sporting event, however, the husband distracted by television seemed more acceptable than the ham operator because of the television viewer's limited involvement. Radio hobbyists took part in two-way communication with real people. In addition to expressing jealousy about the distant men engaged in deep conversation with their husbands, wives made frequent jealous references in the hobby literature to the equipment that was the object of so much tinkering. The ham culture's appeal to technical fraternity and technical interactivity as constitutive of masculinity broke from the norm. According to mid century, middle class standards, men properly displayed masculinity in the domestic context through their relationship to women.[16] A masculinity based instead in technology offended these sensibilities by replacing women with men and machines in a devotional relationship that carried sexual overtones.

In an indirect fashion, complaints that radio reduced spousal companionship reinforced hobbyists' heterosexual identity. The technical fraternity of radio was a circle of men, using a mode of communication linked to covert activity, engaging in private discussions—an intense homosocial network that easily could have provoked suspicion. Succinctly capturing the sexual tension that had surrounded the mid century hobby, a character in a 1992 novel asked, "What do you think those ham-radio buffs really talked about? Do you think some of them were secretly gay, and they left their wives asleep and crept down to their finished basements in the middle of the night to have long conversations with *friends* in New Zealand or wherever?" Except in gender-crossing jokes played under the cover of Morse code, the ham community did not openly question that the hobby existed within a strictly heterosexual environment.[17] But hams' posturing about sexuality displays an awareness that outsiders doubted hams' heterosexuality. Remarks about domestic tensions by hobbyists and their wives confirmed hams' heterosexuality by mentioning their marital status, a preemptive defense against accusations that the level of fraternization in amateur radio crossed a critical threshold. It was better to appear henpecked than face a more pointed charge.

The scene painted by one ham's wife blamed a loss of togetherness on the hobbyist's fascination with radio equipment. Leisure time spent apart made Polly Oltion unhappy that "Hubby was growing less and less familiar." Previously the couple had shared pastimes of discussing world affairs and playing chess, but her husband's solo ham radio pursuits left Oltion isolated. She grew particularly "annoyed and bewildered" after realizing that he used the word "we" to refer to himself and "that disreputable conglomeration of tubes, wires, cans, and noise." Oltion felt replaced as the object of "his affection" by the machinery her husband considered "his bosom pal." While he tinkered in his radio station, she knitted fourteen pairs of socks "to pass away the time."[18]

In humorous descriptions of life with a ham, women wrote of losing out to hobby technology in the battle for men's attention. Nancy Anderson told of a husband who brought his friends right into his hobby area, located in the couple's bedroom, while his wife slept. "The crowning indignity," according to Anderson, was not the breach of privacy. Though there were "more 'hams' in her bed chamber than in a Virginia smokehouse," what exasperated Anderson was that the hobbyists completely ignored the female body. "The lads are so taken up with their wires and tubes they don't even realize the lady's there. Honestly, girls, how much can a woman stand?" Ann Gordon reported being similarly spurned during a courting experience she ironically dubbed her "romantic introduction to ham radio." After Gordon and her suitor had driven up into the hills where they watched "the lights of the city below us, full moon above us," Gordon recalled that she was "beginning to feel in the spirit of things—when out came the microphone and on went the switches." Instead of taking advantage of the romantic and isolated location for physical intimacy, as Gordon had expected him to, her date initiated a conversation with a distant, male stranger.[19]

That radio communication took precedent over conjugal relations became a common joke in the ham community. One hobbyist's wife said she needed to "get another husband for upstairs use." The postcards Warren Bauer used to confirm his radio contacts depicted a woman wearing lingerie and high heels, provocatively perched on a chair, facing a ham engrossed in an on-air conversation (figure 6.1). "But dear I can't go to bed now," the cartoon hobbyist explained, "I'm talking with," followed by a blank space where Bauer would fill in the name of the recipient.[20] The glamorous woman in the foreground radiated a sensual beauty. In a dark corner, the balding ham captivated by technology seemed asexual. The hobbyist who shrank from human warmth in this image had fetishized radio equipment.

Wives voiced frustration that radio technology disarmed their seductive powers. When Sylvia Frank expressed common annoyances with her husband's hobby—he was always in the basement, only discussed radio, had driven away their friends, spent too much money on equipment, and cluttered the house with electronics—friends told her it could be worse if her husband instead drank, gambled, or had an affair. People of that opinion, Frank responded, clearly did not know any radio enthusiasts. For while "The aforementioned pitfalls may be overcome by talking, coaxing, petting, or any number of other methods," Frank knew from experience that womanly charms could not lure a man away from ham radio. Manufacturers played to this sexual tension by suggesting that expensive radio gifts might stimulate men's affection. A 1953 advertisement for National brand hobby equipment carried a drawing of an elegantly dressed couple. The man had pulled his headphones off with one hand, wrapped the other around the woman's waist, and swept her backward with a kiss (figure 6.2). The advertising copy explained, "It's not her perfume" that attracted him, "it's the National she bought him for Christmas!"[21]

The hobby slang for "wife," it must be noted, was implicitly desexualizing. Built upon the ham abbreviation for a girl or woman, "YL," the term "XYL" literally designated a married woman as a "former young lady." Hams applied "YL" to females of all ages, so the title "former YL" suggested not lost youth as much as lost gender. While female hobbyists accepted the name "YL" proudly, only rarely—and then often ironically—did they call themselves "XYLs." One reason women hams reported liking the term "YL" was that it seemed a fitting analog to "OM" or "old man," the slang for a male hobbyist of any age. The well-established association of maturity with masculinity in the phrase "old man" further amplified the insinuation that a change of status from YL to XYL diminished femininity.

Figure 6.1
Warren Bauer joked on his radio confirmation postcards that he was so devoted to the hobby that even a scantily clad seductress could not tear him away from a ham conversation. Bauer had filled in *CQ*'s name on this card when he entered it in a contest held by the magazine. Published in "*CQ* QSL Contest," *CQ*, August 1956, page 63. QSL reprinted courtesy of Barbara Bauer Lawrence.

Figure 6.2

If feminine charms such as perfume could not win a hobbyist's affections, one manufacturer suggested that a gift of expensive radio equipment might do the trick. But the ham in this advertisement remained tethered to his radio by the headphones. National Co. advertisement, *CQ*, December 1953, inside back cover.

Despite frequent illustrations of spousal conflict brought on by the hobby, marriage was the norm in the ham community. One might expect that single men reading about marital woes in radio magazines chose to remain bachelors, reveling in the freedom to allow a hobby to dominate leisure time and the entire home. Instead, many hams sought wives who accepted the hobby, ideally someone equally intent on spending time and money on the hobby. In hopes of appealing to the rare single women readers, bachelors wrote letters describing themselves to hobby publications. But a radio license was not a prerequisite for a potential wife. Hams also favorably considered as partners dates who patiently listened to radio tales and sat through demonstrations. It might be possible, hams wrote, to convert these women into fellow hobbyists later.

Male hams anticipated that being married to another radio operator would eliminate "accusations like 'You think more of those stupid old knobs and dials than you do of your own family'" and "dirty looks when you present her with a nice low pass filter for her birthday." In describing one member's station, a club newsletter called his hobbyist-wife a desirable accessory. Although the club member did "not have the fanciest setup we have seen," the newsletter pointed out that "he has something of which few of us can boast, an XYL who is a ham." Hobbyists who lacked this component could try to make their own. Amid articles that described how to convert surplus military communications equipment to civilian use, *CQ* magazine published "Converting the XYL: New Conversion Data on a Widely-Popular Non-Surplus Item" on how to change wives into hams. The author, Florence Collins, had experienced the process firsthand and told men they, too, could relieve household tension caused by the hobby. "The schematic for a slick conversion job," she promised, would produce "an XYL ham operator to share your enthusiasm for this fascinating hobby."[22]

Hams who did share the hobby with their spouses found that the supposed solution came with its own problems. Florence Collins's husband, James, challenged her conversion instructions with a rebuttal titled "Nothing... But the Facts." True, James said, he no longer needed to justify spending money on radio equipment. But because his wife planned to use the equipment, she wanted to be involved in selecting it. And when the new rig arrived, James had to compete with Florence for air time. Having his bigger hobby budget subjected to compromises made James wonder whether he really was so fortunate to live with another ham. Maude Phillips admitted that after joining her husband in the hobby, she only granted him access to their station when "some ham wants to talk 'technics.'"[23] Nevertheless, if wives participated in the hobby, a large measure of the social and spatial distance created by ham radio disappeared.

Women's clubs that paralleled radio clubs extended to wives what one called an "honorary or auxiliary or associate membership" in the ham community. The Ladies Auxiliary of the Rochester Amateur Radio Association, La-RaRa for short, welcomed "any ladies interested in radio, socially or technically." La-RaRa provided "an opportunity for all ladies to get better acquainted with their boy friends' or hubbys' hobby." Women's auxiliary clubs did not discourage the practice of ham radio, but their primary function lay in supporting the associated men's radio club. In 1952, only five of the thirty La-RaRa members held radio licenses. One Ladies Auxiliary project involved "obtaining neck ties for the men and putting their call letters on them." La-RaRa also catered the Rochester Amateur Radio Association's meetings and weekend-long contests.[24] Such activities kept women engaged with stereotypically feminine tasks while physically near partners who were focused on technical hobby activities.

As a concession to the wives and children ostracized by men's radio hobby, many fraternal ham clubs hosted family gatherings. "Ladies' Nights," "'No-Speech' Dinner Dances," "Family Dinner Meetings," and summer picnics featured entertainment described as "strictly non-technical (for the benefit of the ladies)" and door prizes chosen to appeal to "the fairer sex." The announcements for these functions acknowledged spousal conflict as a normal side effect of ham radio. The Northern California DX Club newsletter reasoned that "The XYL [wife] will surely take a more kindly view of the monthly club meetings if you treat her to an evening such as" the "special social meeting" that was to be conducted at a local winery.[25] Permitting occasional, carefully orchestrated family visits into radio clubs clarified that a boundary normally existed between these social realms.

Though auxiliary clubs for hams' wives and invitations to family events helped ease household tensions about the hobby, numerous disputes about radio lingered and caused bad publicity. Hams struggled to change the hobby's reputation for being incompatible with domestic life. In 1955, the editor of *CQ* magazine sought contributions for a press release designed to make people see ham radio as family friendly. He particularly solicited news of "anyone [who had] managed to build a ham rig into a modern home and keep it unobtrusive." Evidently *CQ* could not gather enough positive examples: the editor repeated the call for happy "ham families" the following year.[26]

Twenty years later, the conflicts hobby radio caused with home life drew less attention. *Ham Radio Horizons* still felt the need to inform readers in 1977 that families would not "resent the time that you spend" on the hobby as much if they understood it better.[27] Electrical interference and the drain on scarce resources continued to annoy those who did not participate in ham radio, but prevailing norms had shifted. No longer did popular culture's idealization of family togetherness subject men's leisure to intense scrutiny. In the 1960s and 1970s, women made fewer public accusations that radio disrupted family life, and some wives spoke out in defense of the hobby. Husbands gained freedom, as individuals, to spend time and money on ham radio.

Men's increased freedom of identity in the 1960s and 1970s is the less-remarked-upon half of a better-known story. Women's liberation contributed to relaxed attitudes toward men's hobbies. The greater number of middle class women working outside of the home and the ideological boost of the women's movement led wives and mothers to take on additional identities. Married women decreasingly defined themselves in relationship to their husbands as they gained power outside of the domestic sphere.[28] And mores shifted in such a way that the nuclear family no longer was the only acceptable basis for gender identity. Hobby time passed separately then appeared less threatening to marriages and to sexuality. One ham's wife in 1979 explained that she tolerated his "mechanical mistress" in exchange for "the same respect" and noted that her husband occasionally took a turn doing laundry or washing dishes so she could "pursue one of my interests."[29]

Evidence from the hobby literature indicates that men's ham radio activity gained domestic acceptance in the 1960s and 1970s. Magazines no longer carried articles blaming hobbyists for neglecting their families, and club newsletters' casual jokes about spousal squabbles fell to a minimum. After periodically publishing wives' complaints about ham radio since its debut in 1945, *CQ* printed no critique of this type after 1960. Then three articles by women in *CQ* during the 1970s reversed the castigating stance of the 1950s and called for wives to support their husbands' participation in the hobby. Although shifting editorial policy must be acknowledged as one possible explanation, the larger framework of gender relations suggests that these publication changes reflected a new context for radio in the home.

The overriding message from the women who wrote for *CQ* in the 1970s was that wives needed to take a kinder view of ham radio. Charlene Knadle compiled a list of questions to determine potential spouses' "Amateur Radio Marriage Quotient." She assumed that in "an average amateur-radio couple, the husband is a ham and the wife either is not or is a less active one." Knadle believed it nearly impossible for any pair to live so peacefully with radio that they would score 35 or more out of the 40 points possible in her quiz. But she held a ham husband and nonparticipating wife equally responsible for coping with the hobby and drew her conclusions about how ham radio would affect a relationship based on both partners' attitudes and behavior.[30]

Other articles went further and placed the greater burden for tranquility in a ham household on the wife. "Just Hams" introduced the fictional Barbara who realized, as soon as she took the time to learn more, what great people hams were and how wonderful a pastime radio was. Barbara faulted herself for having previously "crippled" her marriage with complaints about her husband's hobby. Gail Steckler wrote of a similar change of heart. "Before I understood my husband I was jealous of the time and money spent on the radio," she recalled in 1979. But Steckler had since gained "a healthy respect for" her husband's "individuality" and for "radio as an aspect of my husband's life which happens to be apart from me." She considered it important for spouses to grant each other "time and space alone" and particularly stated that independent male identity could benefit a couple. "When he emerges from his world of radio," Steckler witnessed in her husband "an inner peace and an acceptance of his self which translates into a better balanced relationship for us."[31]

After years of shouldering the blame for disrupting home life, ham radio operators started highlighting the part played by wives. A "message to every Ham's wife (Bless em)" printed in a 1972 club newsletter claimed that "Your husband's Amateur Radio activity can be one of the most important ingredients in your marriage. Whether it works for or against it is STRICTLY YOUR OWN CHOICE to make." "A smart wife," according to the author, "will use her husband's Amateur Radio interest as a primary means of develeping [*sic*] a stable, well adjusted, smiling homelife." To gain the support of wives for ham radio, men were told to "Do your part; be reasonable, meet the XYL halfway." This extended to treating a wife as "an individual too with ideas to be expressed." When the author further suggested that a husband should "Be sure to be interested when she wants 'equal time' to tell you of her day," he indicated the novelty of women's demands for equality by setting off "equal time" in quotation marks. Gradually the hobby literature adopted a more cooperative approach so that by 1979 it was not unusual for a ham manual to advise newcomers to the hobby to "go about it with consideration for your other responsibilities" in order to "fit ham radio into the scheme of your life with the least possible disruption."[32]

Accounts from the 1970s of tolerance and even support of a husband's ham activities broke dramatically in tone with, for instance, Ruth Johnson's 1946 "Wife's Eye View" that the hobby strained marriages and her advice to "prospective wives" to avoid hams as husbands.[33] The hobby of radio was basically unchanged over the three decades, and there is little reason to believe that it caused fewer intrusions. But changes in the domestic sphere meant that adult hams' pastime was perceived differently and conflicts about radio were represented differently. Hobbyists and nonhobbyists still shared the understanding that recreational radio was bound to set a ham apart from the family and to interfere with home life. It was just that social dynamics had shifted in a way that accommodated the freedom men found via ham radio. Separation from the household no longer threatened to call hobbyists' sexuality into question in the 1960s and 1970s, and hams and their wives could relax the rhetoric about marital tensions that had functioned to make hams' independence innocuous by defining it with respect to family units.

Spatial Solutions

No matter what a family's attitude toward the hobby, ham radio posed a fundamental practical problem: where to put the station. At the very least, every operator needed a transmitter and receiver, a telegraph key or a microphone, and headphones or a speaker. The typical hobbyist's equipment list extended beyond these basics to include tools, testing instruments, myriad accessories, and sometimes an extra transmitter or receiver optimized for communicating on a particular frequency band. Add to this log books, maps, wave propagation charts, technical manuals, and so on, and the challenge quickly becomes clear.

Ham radio shacks grew out of negotiations about how to fit a technical hobby into the household. On the surface, this spatial solution addressed the question of locating the radio setup. But specialized hobby areas accomplished something socially for hams as well. The privacy of shacks signaled hobbyists' membership in a community defined outside of the home, facilitating hams' development of identities apart from family roles. Separate space granted to ham radio within the household physically represented the tension inherent in the partial domestication of radio as a home-based hobby employing high-powered, militaristic equipment for worldwide communication. And the gendered domestic architecture that framed hams' search for space substantiated the connection between masculinity and radio technology.

Hobbyists and their families agreed that radio needed its own territory, even when they pointed to different merits of the arrangement. Transmitters that interfered with television pictures, staticky conversations piped in through hams' speakers, and workbenches covered with tangles of radio equipment made stations unwelcome in shared rooms. Those not involved with the hobby saw shacks as a straightforward way to distance themselves from such nuisances, with slight variability in their perspectives. Some mothers like Joy Freed, who thought her son's proud display of his messy hobby area was "a constant embarrassment to all the family" and worried that it discredited her as a housekeeper, took comfort in thinking of a boy's bedroom filled with gadgets as an educational training site. Because establishing shacks soothed domestic disputes by containing ham radio, views of shacks did not vary significantly whether family members were irritated

Figure 6.3
Except for the style of equipment, the typical shack varied little over the decades. Compare this photograph of a shack in 1925 to that of a 1968 shack in the Prologue (figure P.1). Photograph from *Popular Radio,* October 1925, page 307.

by or comfortable with the hobby. Nancy Anderson's concession—amid an otherwise scathing criticism of the hobby written in 1956—that ham radio "may be bearable if the ham has his rig in a shack removed from family living quarters" was similar to the remarks of a ham's wife in the early 1970s who attributed her acceptance of the hobby to the fact that she could "put all the equipment up in one room and shut the door."[34]

To hobbyists, shacks were retreats. Hams turned the problem of situating radio equipment to their advantage and embraced the opportunity to spend leisure time privately in a separate room or just in one portion of the basement or garage. In these personal havens, hobbyists could escape job and family responsibilities and talk with men around the world. One ham's wife described witnessing his liberation when, "After a busy day in a large impersonal office," he "put on his old jeans and disappear[ed] into his shack where he can work by himself and for himself."[35] Technical culture flourished in ham shacks, with radio needs and aesthetics governing all design decisions. As a result, the hobby community considered the shack a material expression of a ham's devotion to radio.

Ham shacks need to be understood with respect to the gendered division of the home for hobbies and other functions. Many leisure pursuits required specialized household spaces, and women commonly controlled such access. As part of the increased focus on children in the 1940s, families attempted to accommodate a greater variety of recreational activities. In *American Home* magazine in 1943, Constance Foster wrote with an awareness that "The new psychology said that children's developing interests were more important than furniture." Still, she had grown weary of clearing away her children's paints and musical instruments "to make the room respectable" every time she entertained guests. Foster's allocation of space in her seven-room house for the ten hobbies of her husband and three children precisely outlined how even women in active families could insist that "The Living Room Belongs to Mother!"[36]

Women's power to assign household territory contributed to the sense that postwar homes had a feminine feeling overall. Constance Foster limited her husband's and children's hobbies to the private spaces of their home. Her daughter painted at an easel in a drafty enclosed porch, and her husband puttered in his woodshop in the unfinished basement. Meanwhile, Foster retained the most refined rooms as her own. The public areas of the house were labeled feminine, and the private ones fell under matriarchal control.

Men supposedly had trouble feeling comfortable in the shared parts of a home designed by women. In 1967, *McCall's* promised that "Nothing is more surely calculated to delight a husband than a room or area decorated to reflect his very special tastes and interest." Wives were told to consider providing husbands with "A Room of his Own," "a little haven to which he can occasionally withdraw and bask in lordly comfort." The sophisticated areas depicted were for looks more than activity. Dark wood, leather-bound books, huntsman prints, clocks, and nautical knickknacks, *McCall's* claimed, would help ease marital tensions caused by blurred gender roles, a problem reported on elsewhere in the same issue of the magazine.[37]

In contrast to the proper rooms of the house, rugged areas like basements, garages, and attics had a manly feel. Gaston Bachelard's 1957 study of how people experience domestic architecture, *The Poetics of Space*, described basements as sites of rationality and practicality. These bleak underground spaces offered an outlet to male family members, as John Wright recounted about his father's 1950s hobby workshop. The Wright household fit the pattern where all of the "finished, tidy, respectable spaces" "belonged" to Mrs. Wright, leaving "only the basement and the garage" for Mr. Wright. The cellar in particular "was understood as a masculine space," in a way that connected to Bachelard's characterization of the basement as a work area. "Here," Wright fondly remembered, "a man could get his hands dirty and not worry about making a mess."[38]

Classification of the basement as a masculine work space made it a popular site for men's hobbies that required workbenches and consequently a gathering place for fathers and sons. Wright called his father's workshop "his place and his only," "a masculine refuge in an increasingly feminized household." Because cellars and the workbenches located there "belonged" to men, fathers could dole out parcels of this precious territory to their sons. Radio hobbyist Adrian Weiss approached his father when he needed a place for his first radio shack and successfully "talked my dad into giving me half of the space on one of his workbenches in the basement." A photograph in *The Electronic Experimenter's Manual* showed how the physical isolation of a technical hobby could draw curious sons near to fathers. In the photo a father repairs a piece of apparatus while his young son stands by the edge of the workbench watching intently (figure 6.4).[39]

Basements and garages, male domains designated as locations for machinery, appeared particularly well suited for radio shacks. Ham equipment stood out as more blatantly technical than the many other appliances common in the postwar home—mostly marketed to female consumers—that concealed functioning parts behind sleek cases designed to blend into the domestic environment. The rugged look of ham radios, along with the frequent need to tinker with radio innards, made them seem more like machines than appliances. Basements also had practical advantages as locations for radio stations. Underneath the house it was easy to ground electrical equipment, and basements did not experience rapid temperature

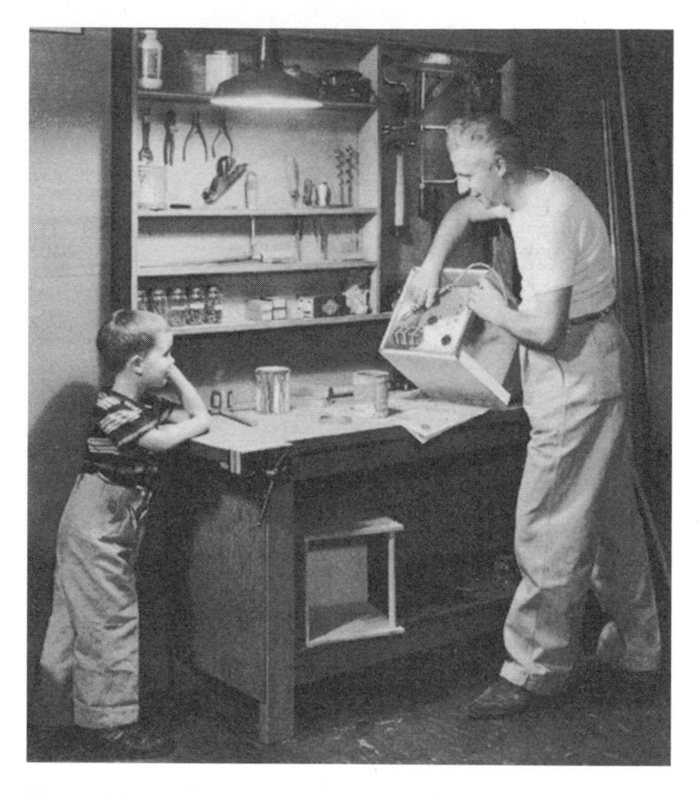

Figure 6.4
The ham radio literature often depicted the hobby workbench as a potential site for father-son bonding. Photograph published in Findlay, *The Electronic Experimenter's Manual* (1959), page 68. Reprinted courtesy of Ziff Davis Media Inc.

fluctuations that could harm sensitive electronics. Even a wealthy bachelor, whose house built entirely around radio was the subject of an article in *House Beautiful*, had chosen to situate his ham shack in the basement.[40]

Boys often were free to operate ham radio from their bedrooms. The articles from *Parents Magazine* about supporting tinkering instructed mothers not to tidy boys' hobby areas. When Joy Freed "began to rebel" at the mess growing in her son's bedroom workshop, she checked her urge to clean it because of lingering guilt from having broken delicate parts during her past efforts. Marianne Besser argued that all boys needed "room at home to work and dream." Parents must set aside their usual authority,

Besser wrote, and instead consider a boy's bedroom as "his place where he's free to do as he likes."[41]

In cases where the radio station occupied a spare bedroom, the hobby took on symbolic status as a family member. The wife of one ham complained that radio gear and the marital bed received equal accommodations. After their wedding, she reported, "We started renting a small two bedroom home, naturally, one for us and one for the ham rig." Family growth meant less room for radio. A new father "got a good chuckle" at a ham club meeting with the story of how his hobby had been squeezed out of the no-longer-spare bedroom. Repeating the member's statement "that he literally screwed himself out of the shack in the house so they made it into a nursery and moved Bill out to one corner of the garage," the newsletter editor concluded simply, "C'est la vie."[42]

Whatever the location of a ham station, hobbyists viewed separation as a critical characteristic. Almost everyone who wrote about establishing a shack pined for a distinct, private space of his own. The American Radio Relay League's *Radio Amateur's Handbook* declared "the amateur with a separate room that he can devote to his amateur station" to be "fortunate indeed." Luckier still were "the few who can have a special small building separate from the main house." The basement or attic, according to the handbook, provided adequate separation, "although it may lack the 'finish' of a normal room." One ham willing to accept any private space for a shack wrote that "The corner of the bedroom, the old coal bin or attic hideaway that contains our station is our own personal pride and joy."[43]

The isolation of the ham radio shack allowed it to be more than space for a hobby. Virginia Woolf focused on "the urbanity, the geniality, the dignity" that followed from "luxury and privacy and space" when she made the case for each woman writer's needs in *A Room of One's Own*. With comments such as "A man's ham shack is his castle!" hobbyists pointed instead to the link between operating two-way radio and acquiring masculine space in the home. "Perhaps the ultimate satisfaction for a radio amateur," according to one, was "to have a space entirely his own."[44] In this regard, the perceived security threat of open radio communication benefited hams by providing a justification for tight control of stations. Woolf's prescription of "a room with a lock on the door" for genuine privacy neatly meshed with the Federal Communications Commission's Cold War regulations.

During the 1950s, radio license applications required hobbyists to pledge that "the station will be under my exclusive control" and "the equipment will be inaccessible to unauthorized persons." A quarter century later, ham radio handbooks still advised storing transmitters "in a locked room, cabinet, or closet" to avoid illegitimate use.[45]

Efforts to distinguish the ham shack from the family home often surpassed the basic needs to secure the transmitter away from the television and confine radio gear to one area. A photograph in a hobby handbook showed the combination garage/shack designed by one ham. Detached from the house, the building offered separation and presumably some privacy in and of itself. But the ham further had reduced access to his leisure space by installing, side by side, separate exterior doors to the garage and to the shack. Another ham found it challenging to keep in touch with the rest of the household while ensconced in his isolated station. Ed Marriner published schematics for constructing an intercom that would allow hobbyists communicating with people around the globe also to communicate with their own families. "Practically any married Ham will appreciate the necessity of an intercom when the Ham shack is in a remote part of the house, or even out in the garage, etc.," Marriner explained. "After all, the call to chow is pretty important."[46]

Finances and other practical limitations prevented most hams from putting hobby needs first when deciding to purchase or build a house. Only a privileged few could obtain the ideal ham radio home—elevated high enough to send and receive signals clearly, isolated from neighbors to minimize complaints about television interference, and free from zoning restrictions on antenna installations. Yet ham publications indicated a clear preference for single-family houses and assumed that most amateur radio operators could afford them. Soon after the war, a hobbyist described a station in a closet as a "clean-cut solution to today's housing shortage." Later some handbooks did mention adaptations of shacks to apartment living, but comparisons between the "full-size workshop" and the "apartment workshop" implied that the latter was a distinctly second-class accommodation. One manual insisted that apartment-dwelling hams must prioritize, purge closets of trivial things, and "Tuck the pajamas under your pillow, and toss the tux under the bed!" The American Radio Relay League described radio as a "democratic hobby" in which technical ability and proper behavior mattered more "than thousands of dollars invested in special equipment and an elaborate 'shack.'" When the League's handbook then proceeded to instruct readers on elaborate shack assembly, though, it cast doubt on the sincerity of this statement.[47]

Ham radio operators who did not have the luxury of practicing the hobby in a private part of the home struggled to cope in public space. Only "the true experimenter," according to one handbook, possessed the focus and dedication to "put up with" the compromises necessary if trying to use a central site like the kitchen table as a workbench. One ham complained his station was "squeezed for space" because it shared the garage with his wife's ceramics kiln and supplies. Another noted contemporary architects' view that recreation rooms "should be a center for the spare-time activities of the entire family," but he tried to shield himself from the hubbub of leisure pursued in a group. He argued that the "Post-War Ham Shack" could serve "double duty" within the household just as well if it were established in a rarely used guest bedroom. When the station had to share space with the whole family, hams sought privacy by staggering their activities. They pointed out, for instance, that it was convenient to chat with hams in distant time zones while the rest of the family slept. Otherwise, radio operators at least could shield the incoming half of their conversations by wearing headphones.[48]

Hobbyists with stations in common areas of the home concealed radio equipment or altered its appearance to meet family aesthetics. Nicholas Lefor described a living-room ham station stowed inside a Sears Roebuck steel wardrobe cabinet. "In adapting this cabinet to an amateur transmitter," Lefor explained, "appearance was the prime consideration." A centrally housed station carried the risk of family members meddling. Lefor cautioned readers to "make sure your wife does not use this cabinet for the purpose it was originally intended, as was the case here, where a few pairs of shoes were found on the bottom shelf." To appease a wife characterized as "lord and master of the arrangement of furniture and the overall appearance of the home," a hobbyist who lived in a three-room apartment published a plan for refitting a tall secretary-style desk to hold radio equipment. He called this "The Good Housekeeping Approach to Station Design."[49]

Those with the freedom to tailor spaces to ham radio took this as a serious responsibility. One hobbyist, expressing reverence for "better than ever" postwar communication equipment, advised hams not to "make the mistake of housing it in a 'shack' which is so inadequate that it will spoil half your fun." Thirty-five years later, Dave Ingram claimed that the right hobby atmosphere could "add 3 to 6 dB's [decibels] to your 'DXing confidence'" by providing inspiration during long contests or in periods of heavy activity. A thorough approach to shack design extended to wall decorations. Ingram regarded "DX awards, current propagation charts, maps and photos of distant lands one has contacted" as "far more *DX-inspiring* than family reunion photos and model airplane collections." In a chapter on "Assembling a Station," *The Radio Amateur's Handbook* half-observed and half-commanded that "most amateurs take pride in the arrangement of their stations, in the same way that they are careful of the appearance and arrangement of anything else that is part of the household." Howard Pyle, author of another ham handbook, likewise advocated attention to style and detail in the station. If a plywood panel served as an equipment base, for instance, Pyle indicated that it should be "covered with linoleum, micarta or similar material" and that the edges should be trimmed "with chrome molding for best appearance." He recommended that hobbyists who planned to spend "many pleasant hours" in their shacks "make the surroundings attractive as well" as functional.[50]

Of course, practical considerations specific to the hobby also entered into shack design. An equipment manufacturer polled hams in 1956 and found "the ultimate desire of all was to have equipment which 'went together.'" The concern here was not that colors match, but rather that cabinets be physically compatible. Results of the survey revealed that "The difficulty of installing odd sizes of cabinets has always been a source of irritation to the neat and efficient operator." Handbook chapters on setting up a station addressed the arrangement of nonuniform equipment and assorted other "workbench tricks." Novice hobbyists relying on these guides learned, for example, that muffin tins and ice cube trays were handy containers for keeping small parts organized and that mounting a voltmeter on a tilted rack made its face easier to read.[51]

Hobbyists felt they could get to know a ham by viewing his shack. The newsletter of the Northern California DX Club was one of many club publications to introduce members by way of their personalized radio environments. Beginning in 1948, *The DXer* featured a monthly column picturing a ham posed in his home station. The limited circulation and budget of the newsletter meant that a printed photograph had to be pasted into each copy by hand. This process must have been tedious, yet the club retained the shack profile feature until the mid 1960s.[52] Hobbyists' desire to document their shacks "to show to other hams, to send to radio magazines, or to show 'before' and 'after'" led *CQ* magazine to publish articles with tips on photographing the typically cramped spaces.[53]

Station design and day-to-day tidiness were important aspects of a hobbyist's image. Visitors appeared often enough that handbooks advised readers to plan for extra seating in shacks. For fellow hams, a visit offered a view into the "personal preferences and life style" of the shack's inhabitant. For those outside the ham community, a visit to a shack had the potential to shape their opinion of the entire hobby. A 1956 article suggested hams "Clean Up the Shack" to avoid the need "to apologize for the appearance whenever a visitor comes in." A decade later, a very similar article put the burden on each ham "to demonstrate the worth-while activities of your hobby to your visiting public if you want to help to keep such hobby alive." Ham radio's very survival, by this logic, depended on "a neat, clean and efficient operating center."[54]

Shacks grounded the ethereal hobby experience and situated hobbyists in a domestic context. In these ways, shacks contained and partly domesticated a worldly technology. In other ways, shacks differentiated radio technology from the household. Hams occupying separate spaces visibly marked themselves off from the rest of the family, at least during leisure time. The technical nature of ham radio alone did not justify isolating stations. At mid century, many new technologies were integrated into households' existing physical and social structures. It was the peculiarity and roughness of hobby gear that made shacks appealing to non-hobbyists. Based on those same characteristics, radio typically received space in unrefined, manly parts of the home. The hobbyists who took refuge there did not mind the conditions. Shacks turned two-way radio's incompatibility with home life into a source of freedom.

Radio hobbyists developed a technical identity that operated in a tense but constant relationship with home life to grant them independence.

Out of debates about the allocation of household space, time, and money to radio, hobbyists gained a stronger sense of belonging to a community apart from the family, with its own values and practices. Public recounting of the social stress caused by radio reminded all that hams indeed had wives and children, while establishing the identity of "ham" as distinct from that of "husband" or "father." Practicing the hobby functioned as a retreat from the household and the roles and responsibilities typical of it. It was the hobby self—not the husband or father—who talked on the air with other hams. "Freer men" emerged when radio interrupted domestic relationships, just as the accumulation of odd equipment in the corner of a room delineated a shack.

7 Technical Change and Technical Culture

Integrated circuits radically changed electronics. Even when introduced in the early 1960s, single integrated circuits (ICs) could perform the functions of dozens of previous-generation components such as transistors, resistors, and capacitors. At an astonishing rate, electronics engineers increased the capability and decreased the size of ICs. Soon tiny chips of semiconducting material could perform hundreds of operations. Compact designs for devices dependent on electronics followed accordingly. Computers that once filled rooms shrank to fit onto desktops, then onto laps, and into palms.

For the consumer electronics industry, integrated circuitry was a boon. The substantially lower costs to manufacture IC-based products more than made up for the expense of reeducating workers and retooling factories. Demand accelerated as customers replaced entirely serviceable electronic devices with smaller versions that they found more convenient or simply intriguingly hi-tech. During the industry-wide adoption of ICs in the mid 1960s, manufacturers of hobby equipment and kits continued the strategy that had served them well for decades—they accommodated the small ham market by taking advantage of production flexibility, building ham radios from the same stock used to make other electronics equipment. But this strategy failed with the latest state-of-the-art components. Including ICs in ham radio products altered the hobby market and threatened the technical culture of hobby radio.

Integrated circuits made hams reevaluate the merits of building equipment and buying ready-made gear. For several reasons, home construction became less appealing. Its cost advantage diminished as prices fell for commercial products made from ICs. Hobbyists using standard tools and their own hands found it frustrating to work with miniature components and impossible to duplicate the compactness of factory-built equipment. Plus, those fond of tinkering did not always appreciate the labor savings that resulted from replacing multiple transistors with one integrated circuit. Most significant to the ham radio community, building with ICs provided few technical lessons. Component manufacturers, in the same way they had supported amateurs through the shift from tubes to transistors, were "only too happy to flood the enthusiastic IC hobbyist with tons of application information and specifications on their line of experimenter ICs." The guidance extended by the electronics industry was strictly practical, with theory "held to a minimum" because, as a guide from Motorola explained, "a detailed explanation of integrated circuit operation is not necessary to construct these projects."[1] Hams learned to build with ICs, but the chance to apply existing technical skills or gain new knowledge disappeared from building. Integrated circuits were black boxes. Unlike glowing vacuum tubes and visually distinguishable solid-state parts, the ICs wrapped in opaque, standardized shells gave no clues about how they worked or what they did.

The same problems reduced hams' interest in kit assembly. Beginning in the 1970s, kit producers gradually included more integrated circuits to follow trends in ready-made equipment. This unwittingly decreased satisfaction and learning by doing. Following a strong increase in annual sales from $19 million to $28 million in the mid 1960s, the demand for electronics kits fell off in the 1970s.[2] Kits had never overtly taught function or theory. What could be learned from a kit largely was limited to what the assembler inferred from connecting the components. Integrated circuits hid these links, and even the schematic diagrams of kits that contained ICs revealed little about the basic electronics operations underlying the layout. When the Heath Company stopped producing kits in 1992, its president attributed the decline in demand for Heathkits partly to the mystery of integrated circuits. "Now when you drop in one IC, you've dropped in 432 components," he explained, "and the customer has no idea what is going on in there." Kits with complex, sealed components encouraged only clean soldering and attentive following of instructions, not deep understanding. Assembly then felt like a dull routine to one formerly "kit-smitten" hobbyist, compared to what had been a process that allowed "persons inclined to slide rules and horn-rims" to feel creative. Another kit assembler remarked in 1975 that the only occasion for learning by doing came if the device failed to work initially and required tinkering, something he otherwise found absent from "being an electronics buff today."[3] As kits presented fewer opportunities for technical interactivity and only a slight financial saving, they lost value as a middle ground between building and buying.

Many hams trying to stay connected to their technical culture turned to older equipment, which had core community values built into it. Radios that did not contain solid-state parts required frequent tinkering to keep functioning, especially decades after production. Vintage gear also was less predictable, making attempts at routine radio contacts more exciting. And some hams enjoyed that radios built from tubes sounded as they had in fondly remembered years past. (Unlike audiophiles who argued that analog vacuum tubes produced a clarity of sound not possible with digital circuitry, hams did not express an appreciation for ideal audio characteristics, but only for the tone of old transmissions.) Two articles in the fall of 1971—one reviewing a ready-made receiver that had been popular in the early 1930s, the other giving instructions for building a 1930s-style transmitter—later were credited with having "boosted a longtime interest in old-time ham gear." The collection, restoration, and operation of vintage radio equipment became a subhobby of ham radio, with specialty clubs and publications. Poking fun at hobbyists willing to spend extra money for outdated gear, a cartoon showed a ham leaving the "nostalgia alley" section of a used-radio swap meet carrying a piece of equipment with a $300 price tag. His wife, spotting his own call sign already inscribed in the cabinet, remarked, "That looks like the same radio you sold for $25 about fifteen years ago!" (figure 7.1).[4] These hobbyists, who identified with ham radio, seemed also to be seeking to recapture their own pasts through old technology.

More than any other device, the vacuum tube became the object of hams' technological nostalgia. Hobbyists reacted to the proliferation of integrated circuitry with sentimental comments about the pleasures of tubes and the comforts of a bygone period and style of electronics. One noted that "the warm glow of the tubes brought far greater satisfaction than those plain black 'solid state' parts." A former Heathkit employee agreed that there was

Figure 7.1
Hobbyists trying to recapture the past through vintage equipment, this cartoon joked, drove up the prices of the same radios they had abandoned years before. Reprinted with permission from Beasley, *The Best of Beasley* (1994), page 22.

"something particularly thrilling" about constructing earlier electronics kits because "the warm glow of the tubes made those products seem to come alive in comparison with products of the current solid-state generation." Without "the soft glow of a vacuum tube," another amateur went so far as to claim that electronics based on ICs "no longer has a soul" and was "functional but cold."[5]

The emotional attachment of radio hobbyists to the vacuum tube in part denoted a desire to maintain certain community values associated with the technology. Construction projects with tubes, a handbook pointed out, had required hams "to be cautious of high voltage, and be careful not to drop a tube on the floor." ICs did not impose the same level of discipline on hobbyists. In 1968, a ham blamed new technology for altering the hobby culture and complained that "Ham radio just *isn't* what it was years ago and we're all living in a fantasy world trying to build our hobby on the values of yesteryear."[6]

The analyses that most bluntly cited changes in hobby electronics as evidence of social change came after Heathkit production ceased. Characterizing it as "Ending a Do-It-Yourself Era," the *New York Times* made Heath's discontinuation of kits front-page news. A long line of commentators joined the *Times* in linking the reduced demand for kits to a shift in values. In an "instant-gratification society," Heath's president found the declining popularity of Heathkits unsurprising. Former Senator Barry Goldwater, who had assembled more than a hundred of the kits, called the end of production a sign that "people today are getting terribly lazy, and they don't like to do anything they can pay someone else to do." When hobbyist Avery Comarow called Heath headquarters seeking an explanation for the company's decision to stop selling kits, he felt unable to communicate his disappointment across the generation and gender gaps he encountered. "The company spokesperson, a woman, was unemotional," he observed and then concluded, "She was from a different era." Another Heathkit fan expanded on his own belief "that we've come to the end of a technological era" by stating that the skills in "proper assembly" and the attitudes toward problem solving that kits had taught were irrelevant to "today's generation of electronic engineers, technicians, and repair people." To Harry Stine, the discontinuation of Heathkits symbolized the overthrow of his technical values. "Life without Heathkits is going to be different," he wrote, "because Heathkits and what they taught my generation are no longer meaningful."[7]

The poignant outpourings about Heathkits in the early 1990s actually referenced an earlier shift in electronics. Despite each of four published memorialists' supposed affection for Heathkits, none mentioned having completed one recently: two admitted that it had been many years since they had assembled a kit, another exclusively discussed kits from the 1950s, and only the audiophile among them demonstrated any knowledge of Heath's product line as late as the 1980s. The Heathkit fans further dated their laments about technical change by proclaiming a nostalgia for vacuum tubes, though it had been decades since Heath included tubes in its kits.[8] Even in the minds of active electronics hobbyists, kit assembly remained tied to mid century electronics.

Years after sales stopped, Heathkits still have a loyal following whose energy is of necessity directed at maintenance and restoration rather than buying and assembling the latest model. Dedicated Heathkit fans gather on email discussion groups to share information, and they have published histories of Heathkits as well as advice manuals and *The Heathkit Technical Journal* for keeping old equipment operative. Purists hunt down authentic Heathkit parts for making repairs and rush to purchase the unbuilt kits occasionally sold by collectors. Within this network of electronics hobbyists, accuracy to original Heathkit design takes priority over technical performance. Modifications are frowned upon, and kits that were assembled without deviation from Heath's instructions note "no mods" as a feature in advertisements. Attempts to keep true to the original design extend to obsessions with details such as which commercially available spray paint best matches "Heathkit green." Though they discussed all kinds of electronics equipment, the vast majority of contributors in two Heathkit email groups I monitored for several months identified themselves as radio hobbyists by signing notes with their license number or "73," ham code for "best regards."

For a community defined around a technology, technical change represented a threat to community structure. Integration at the level of electronics components enabled an unwelcome social integration, allowing those without technical skills to participate in electronics hobbies. On the cover of its 1965 catalog, Allied Radio advertised "electronics for everyone" (figure 7.2). Such a broad campaign likely offended core customers who saw themselves as members of a technical elite. When Allied Radio encouraged "men and women of all backgrounds and all ages" to assemble electronics kits with the assurance that kits no longer were "specialty items for technicians and ham-radio operators," it diminished the appeal of kits to those who wanted to demonstrate technical skills by completing kit projects, and to do so within a masculine, educated community.[9]

The emotionally charged technological nostalgia of ham radio operators explicitly conveyed an anxiety about the breakdown of the boundaries to their technical fraternity. More subtly, it also resonated with other anxieties. The integration of circuitry coincided with a number of technical and social trends, from the replacement of interactive technology with characterless machines and the automation of manufacturing jobs in the

Figure 7.2
Simplified kits and Citizens' Band radio greatly expanded the accessibility of electronics pastimes in the 1960s. Cover of Allied Radio 1965 catalog. Reprinted with permission.

electronics industry, to the decline of radio hobbyists' technical status and the growing ambiguity of gender roles. Any or all of these consolidations of responsibility and identity may have intensified hams' longings for obsolete devices. The technological nostalgia that emerged in the 1970s remained an element of the culture of ham radio in the late twentieth century and beyond.[10]

The redesign of hobby radio equipment was a minor side note next to the revolution that integrated circuits caused in electronic computing equipment. Along with controlling large-scale computational machines and desktop computers, semiconducting technology in the form of microprocessors supported the operation of all manner of noncomputing devices. Before long the phrase chosen to describe the technical scene shifted from the general "age of electronics" to specify the key electronics technology. "The computer age" had arrived.

The dominance of computers in the national technical culture profoundly weakened the position of ham radio. Computers, not two-way radios, inspired fears of strategic threats and hopes for strategic potential. The greatest opportunities for career advancement through recreational learning by doing came from tinkering with computer hardware or software, not with radio equipment. Hackers replaced hams as the reigning amateur technical pioneers. As a result the elite, manly reputation hams had built upon the image of radio hobbyists as militaristic, innovative masters of technology began to crumble. The only positive effect of this lost stature was that the hobby found easier acceptance in domestic life. In the military and industrial contexts, hams just lost clout. Ham radio became a technical hobby like many others, distinguished only when operators volunteered to provide emergency communication services.

Their drop in position in the technical hierarchy led some hams to speak up for the purely recreational value of the hobby. High-profile NASA activities in particular cast doubt on radio hobbyists' place "in a swiftly moving, exciting technical world." A 1969 editorial in *CQ* magazine directed at complacent hams in "wood-paneled shacks gabbing into equipment which, by space-age standards, was obsolete ten years ago" warned that "the electronics world is fast rushing past amateur radio" and urged hobbyists to "start paying a little more attention to the electronics end of what's going on at the space centers." In response came letters to the editor that advocated leaving hi-tech pursuits to the professionals. "We can't all be technicians, researchers and engineers," wrote one reader. Another pointed out that "for the majority of us, amateur radio is a hobby and a relaxation," and "we should not be expected to chase after the leaders in the science to qualify our existence."[11] Rapid technical innovations forced hams to adopt a humility in sharp contrast with their earlier technical pride.

Delivering a final blow to ham radio's status, Citizens' Band radio brashly made apparent to all that two-way radio communication required neither specialized knowledge nor orderly behavior. The Federal Communications Commission (FCC) established the Citizens' Band (CB) Radio Service in 1945 for communication within large businesses or between central offices and their traveling employees. It was an easy-to-use, low powered, medium range, two-way radio system intended for exchanges between known stations. To reduce the threat raised by open radio communication and the risk of CB causing interference, the FCC banned CB operations across distances of more than 150 miles and allowed adjustments to transmitters to be made only by commercial radio license holders. With the power of the Citizens' Band constrained in this way, the FCC granted a license to every adult citizen of the United States who requested one, without any type of examination. Through the 1950s only a modest number of operators used the Citizens' Band. Then the price of equipment dropped considerably, leading many more people to participate. By 1963 the nearly 450,000 licensed CB operators already outnumbered licensed ham radio operators by nine to five. Enthusiastic users expanded CB from a business tool to a general communication medium and a hobby. Amid the Citizens' Band craze that began in the late 1960s, it was impossible to know how many "CBers" were active because the majority were unlicensed, but estimates put the number as high as 15 to 20 million during the technology's peak popularity in the mid 1970s.[12]

The culture of CB radio was as free as amateur radio's was restrictive. When CB operators gained a reputation for engaging in subversive and illegal activities, they seemed almost to flaunt it as a mark of antiestablishment leanings. Instances of "circumvent[ing] the law" by using CB radio to alert fellow drivers about police speed traps along highways became common after speed limits were reduced to 55 miles per hour in response to the oil crisis that began in 1973. And the news media and popular culture drew attention to the rarer cases of robbery, drag racing, prostitution, narcotics trafficking, and smuggling facilitated by two-way radio.[13] The Citizens' Band in no way encouraged illicit communication, but, like the telephone, it did allow communication by all, for all purposes. Quite the opposite of how the ham radio community set strict standards for the skill level and conduct of members, fans of CB proudly described the lower price of equipment and the absence of technical demands on users as characteristics that democratized the airwaves and created a version of "Two-way Radio for Everyman." In contrast to resolutely apolitical hams, the president of the National Negro Business League in 1976 spoke of the "political potential" of "a CB in black hands," such as for coordinating driving voters to the polls. *Ebony* magazine named "vigilante purposes," especially "during busing and fair housing marches," as "one of the main reasons blacks and whites began using CBs in Louisville, Boston and Chicago."[14]

The homogeneous community of ham radio operators that had grown out of a particular culture of two-way radio technology reacted defensively to the community of CB radio operators, which was integrated across race, gender, education, and class lines. Frustrated that "the general public seemed to be confused as to just who was who," hams sought a moral-technical high ground and disparaged CB operators as "rule breakers" and "10-4 maniacs." Hams mainly spoke of wanting to limit their community to those with technical skills, yet technical skill was closely associated with certain socioeconomic categories and, in the minds of hams, with certain attitudes and behaviors. While the editor of *CQ* magazine partly supported easing amateur license requirements to accommodate the "many electronics technicians, engineers and other technically-oriented fellows" who were unable "to master the [Morse] code," he hesitated because it would be difficult "to differentiate between the technician/engineer and the CB'er that even the CB'ers don't want."[15] The technical boundary to the ham radio community drawn in such comments signified a host of social boundaries.

The combination of the arrival of the computer age and the accessibility of CB radio at first dramatically dampened enthusiasm for ham radio. From 1965 to 1975—roughly corresponding to the decade when the public took up CB radio for general communication and when mass-produced integrated circuits spread out from the most cutting-edge aeronautics applications to reach consumer electronics—the number of licensed ham operators increased at the slowest rate in the hobby's history, creeping up just 0.5% over ten years and suffering its only period ever of consecutive years of decline (1972–1974). Then a rebound occurred as the ham radio community grew by 40% over the next five years, perhaps gathering internal strength through opposing a rival culture of recreational two-way radio. Meanwhile, a fluke of nature and a regulatory decision brought about the sudden collapse of the CB fad. Increased sun spot activity in the late 1970s, which improved opportunities for long-range ham radio contacts, made short-range CB conditions unpredictable. In 1976 the FCC nearly doubled the number of channels allotted to the Citizens' Band. This eased crowding, but not in the way planned: the new transceivers necessary to take advantage of the expansion were so expensive that most CB operators held onto their obsolete models. Having outlasted the threat posed by CB radio, the ham radio community consistently grew through the end of the twentieth century. Two-way radio activities of this period, however, failed to capture the spotlight.

In the mid 1970s, technical hobbyists gained a connection to the computer age. Personal computers—like the Altair 8800 sold as a kit in 1975, and the Apple II, the Commodore PET, and Tandy-Radio Shack's TRS-80, all released in 1977—made hardware and software available for recreational tinkering. Curiosity about novel electronics attracted many ham radio operators to computers. Some gave up radio for computing hobbies; others incorporated personal computers into ham radio.

The amateur radio community supported the emerging computer hobby community. As employees of technical firms, hams witnessed large-scale computers in operation years before the debut of personal computers and sometimes obtained access for after-hours experimentation. Their existing social-technical community provided a network for pooling ideas.[16] When Bill Waggoner sought collaborators for "some amateur-oriented projects" to be performed with an IBM mainframe available to him at the University of Connecticut, he put out a call to fellow hams. "I am interested in contacting amateurs who are building computer terminals or doing any kind of computer work, hardware or software," Waggoner wrote in a 1969 letter to *CQ* magazine. The positive response to articles about computers published in ham radio magazines like *73* and *Radio Electronics* inspired Wayne

Green (founder of *73* and a former editor of *CQ*) to create the computer magazines *Byte* and *Kilobaud* in 1975 and 1977. Green directly modeled *Kilobaud* on *73*, describing each "as a medium for hobbyists to contact hobbyists—sort of a large scale newsletter."[17]

Where the communities of radio and computer hobbyists intersected, individuals carried the culture of one group into the other. In soliciting explanatory articles from computer manufacturers for *Kilobaud* with the hope that it might "free us from thinking of the CPU as a black box," Green exhibited a ham's desire for knowledge of the inner workings of electronics.[18] The crossover of hobbyists explains the resemblance of computer and radio hobby clubs and publications. It also may have played a role in shaping computers, with values from the amateur electronics community designed into the functions and form of the first personal computers. Though further research is required to demonstrate this with certainty, evidence from the secondary literature on computer hobbyists speaks in favor of such a connection. Members of the Homebrew Computer Club and others who advocated for the creation of understandable and interactive computing technology often referred to their background in ham radio. The possible influence of ham culture can be heard, for instance, in the belief that "the user's ability to learn about and gain some control over the tool" depended on being able "to spend some amount of time probing around inside the equipment."[19]

One of the many handbooks suggesting how to combine ham radio and personal computer activities claimed that "Rarely has there been a better marriage of two hobbies than this one." The use of computers to streamline, and perhaps to enliven, monotonous paperwork—processing QSL cards and maintaining the ham's log of contacts and records for contest entries—was a common application of word processing and database software in the shack. The hobby literature also described turning tasks at the heart of ham radio over to the computer. Recommended operations ranged from the translation of letters typed on the computer keyboard into Morse code to the "futuristic" possibility of having a computer complete a radio contact "all by itself!" In the fully automated scenario, a handbook assured readers that the hobbyist still "would be the control operator, standing by to take over in case the computer developed some Novice tendencies."[20]

Whether experimenting with computers or just using them to update radio communication, interfacing the two kinds of electronics let late twentieth century hams engage with new technology. A ham radio club in Rochester, New York, invited an employee of a local computer company who also was "a computer hobbyist with an interest in voice synthesis" to present "ham shack computer uses with a new twist" at a 1982 meeting. In a period when "modern radio communication is so dependable—and the hardware leaves the operator with little to do but *talk*," and when personal computers were less predictable, one handbook explained that "in the merger of microcomputers and amateur radio, thousands of hams have rediscovered some of the pioneering spirit of a bygone era." Involvement with contemporary hi-tech devices allowed hams to make a case for the relevance of their recreational activities to jobs in the computer age, and the hobby literature of the 1980s included data control and processing on the list of skills developed through ham radio.[21]

Hobby radio survives into the age of the Internet. Rather than shun the latest modes of communication, hams have established news groups and Web sites devoted to amateur radio. Participants in these virtual communities typically represent themselves both with their radio license call signs and with Internet identifiers such as their email addresses or locations of their personal Web pages. The layering of one communications technology atop another seen in hobbyists' simultaneous use of the Internet and radio can appear quite ironic. When QSL "cards" are emailed to save time and postage, an electronic message sent via Internet is accepted as confirmation of an electronic message sent via radio. And enthusiasm for radio activities generates conversations online that earlier would have taken place on the air.

Somewhat surprisingly, ham radio has continued to gain in popularity into the first decade of the twenty-first century. While I have not done the sociological investigation of contemporary hobbyists that would be necessary to explain this trend, I can point to two contributing factors. Very many of the technical barriers, including knowledge of Morse code, have fallen away from daily participation as well as from the initial hurdle of the licensing examination. There is still a core of electronics experts active in ham radio. They are joined, however, by some hobbyists who use the amateur bands as if they were just versions of the Citizens' Band. Ham radio additionally attracts people who appreciate its unpredictability. They perceive the occasional challenges to communication as enhancing the thrill of completing audible connections. In this regard, the current practice of ham radio in part functions nostalgically to replicate earlier technical experiences.

The biggest difference in ham radio in the late twentieth century was its drop in stature as a result of lost distinction. The global scale of radio communication previously had separated hams from their families and neighbors—and associated them with the military-industrial complex—by granting hams what at least seemed like access to a wide geopolitical realm. Even for hobbyists who mostly made short-distance contacts, long-distance, person-to-person communication was a possibility, one that did not exist for the public at mid century. Direct-dial international telephoning gradually began to change that in the late 1960s. The dramatic change came with the Internet in the 1990s, when real-time typed discussions and nearly instantaneous email exchanges put strangers and friends around the world in contact without expensive per-minute telephone charges.[22]

The mid century ham community had so strongly linked values and technology that technical change in the 1970s jeopardized its culture. Equipment containing integrated circuits made it difficult to interact with electronics. Removing the moral and technical lessons offered by tubes and transistors, ICs were an affront to the culture of hobbyists. Instinctively, hams reached for equipment from decades past. The image of radio operators as powerful, skilled, precise, and manly had been based, directly or indirectly, on their identification with a particular form of technology. The shift to ICs as the fundamental components of electronics undermined that image. In bemoaning the technical change, radio hobbyists alluded to a host of larger social-technical changes they were enduring. As control and knowledge slipped away from skilled users of technology—in the workplace and at leisure—users lost independence, purpose, pride, and identity. Multiple distinct electronics parts were integrated into featureless units, and the men who identified with electronics felt this technical change socially. Nostalgia for older technology equally expressed a nostalgia for older values, as hams saw the two bound together.

Ham radio operators may have more directly encountered the integration of circuitry and reacted more strongly to it, but the general population also reflected upon this technical change. In 1965, Thomas Pynchon indicated the diffusion of circuit culture when Oedipa Maas, the heroine of his novel *The Crying of Lot 49*, views the city of San Narciso as a circuit board. Looking "onto a vast sprawl of houses which had grown up all together, like a well-tended crop, [...] she thought of the time she'd opened a transistor radio to replace a battery and seen her first printed circuit." The connection between cityscape and circuit board goes beyond their neatly ordered networks. Oedipa observes San Narciso as a bland composite, "less an identifiable city than a grouping of concepts." Initially the city is indecipherable to her because of its standard appearance: "if there was any vital difference between it and the rest of Southern California, it was invisible on first glance." Oedipa feels, too, that city and circuit share "a hieroglyphic sense of concealed meaning, of an intent to communicate. There'd seemed no limit to what the printed circuit could have told her (if she had tried to find out); so in her first minute of San Narciso, a revelation also trembled just past the threshold of her understanding."[23] Circuit boards in the mid 1960s—as did integrated circuits a few years later—typified slick hi-tech objects, appreciated as packed with information even when the layperson could not comprehend just what that information was. Electronics circuits seeped into consciousness. The public had a notion of what circuit technology was and used circuit technology as a metaphor to put personal experience into words. Said another way, people identified circuits and identified with circuits: there was a circuit culture.

Technical culture emerges from the two interrelated processes of technical identification, creating meanings for technology and perceiving self in relationship to technology. The culture of ham radio provided an exceptional example. Hams developed a technical culture around two-way radio in a way that delineated a community. Through ham radio, men found solidarity with other like-minded individuals, gained access to the kind of hi-tech gadgets that fascinated them, and won the respect of technical employers, including the military. Participants became known as hams and, even more influentially, as technical masters in a society that celebrated technical achievements. These clear consequences made ham radio a good case for explaining the concept of technical culture in that it was relatively easy to view the dual processes of technical identification in action. The downside of focusing on such a potent technical culture is the risk of inadvertently suggesting that it is representative or a benchmark. On the contrary, I intend the vividness of this exceptional example to raise awareness of obscure examples. Many more subtle technical cultures draw strength from precisely the attributes that make them hard to notice—they are pervasive and naturalized. The consideration of ham radio's technical culture points toward broader questions of how we think about and think with technology. Only after we acknowledge the significance of these social and personal aspects of technology can we make informed, responsible choices about the role of technology in our culture.

Acknowledgments

I am indebted to numerous institutions and individuals who supported this project. Facilitating the completion of the book manuscript, I had the tremendous luxury of being paid to devote months to thinking and writing. The Max Planck Institute for the History of Science granted me a postdoctoral fellowship, and Institute executive director Lorraine Daston welcomed me into her lively "Knowledge and Belief" research group in Berlin. The Art, Science, and Business Program at the Akademie Schloss Solitude outside of Stuttgart and Akademie director Jean-Baptiste Joly provided me with a residency in a peaceful, inspirational setting.

Insightful advice over the years from Lizabeth Cohen, Peter Galison, Edward Jones-Imhotep, David Kaiser, and Trevor Pinch focused my questions and strengthened my analysis. Matthew Jones and the Department of History sponsored my appointment as a visiting scholar at Columbia University for the final stage of work.

Early research was funded by fellowships from the Smithsonian Institution, the Smithsonian's Jerome and Dorothy Lemelson Center for the Study of Invention and Innovation, the Mrs. Giles Whiting Foundation, and the Graduate School of Arts and Sciences at Harvard University. For graciously permitting access to collections of rare ham radio materials, I thank Ed Gable of the Antique Wireless Association, the headquarters staff of the American Radio Relay League, and especially Trudy Maxwell and the late Jim Maxwell.

Friends and family sustained me through the past two years of frequent moves. I am particularly grateful for the warm hospitality I received during this period from Jimena Canales and Bill Rutledge, Joan and Allen Haring,

Lana Hill, Rory Riggs, Marylea and Gianni van Daalen, Mark Wasiuta, and Susan Yarnell.

December 2005
East Hampton

Notes

All quotations retain the emphasis present in the original source. Author, title, and page numbers are stated for all materials in which they appeared. Precise locations of excerpts are indicated only for texts longer than five pages. Dates listed with annual reports refer to the subject year; publication typically occurred in the following year.

Prologue

1. Farman, *Tandy's Money Machine* (1992), 114.

2. This composite draws on fragments of evidence scattered through decades of hobby magazines and newsletters as well as data presented in: Wayne Green, "*CQ*'s Survey," *CQ*, December 1957, 34–35; Stanford Research Institute, *Amateur Radio* (1966); and "1968 *CQ* Reader Survey Report," *CQ*, April 1968, 36–39, 114, 116. The characteristics of the average ham remained very stable over time. Surveys indicate that licensees actually were quite evenly distributed by age despite repeated laments about the "graying of the hobby."

3. Henry G. Elwell Jr., "Amateur Population of the World," *CQ*, April 1960, 57, 112–113.

4. *The Atomic Cafe* (1982).

5. Nancy Anderson, "Hobbies," *CQ*, April 1956, 87–88.

6. S. Douglas, *Inventing American Broadcasting* (1987); DeSoto, *Two Hundred Meters and Down* (1936). Among the general histories of radio, Hilmes's *Radio Voices* (1997) is notable for a discussion of women hobbyists not available elsewhere. The final chapter of Susan Douglas's *Listening In* (1999) offers a summary of ham radio activities across the twentieth century.

7. *The Standard Periodical Directory* (1964). The reported average monthly circulation of *QST* in 1940 equaled three-quarters of the number of licensed hobbyists. *QST*, June 1941, 29.

1 Identifying with Technology, Tinkering with Technical Culture

1. Our Foreword, *The Modelmaker*, December 1925, 178–179.

2. This general description of hobbies is drawn from the thorough historical analysis provided in Gelber, "A Job You Can't Lose" (1991), "Do-It-Yourself" (1997), and *Hobbies* (1999).

3. In particular, I want to avoid an oversight common in the literature on amateur science. There studies have been so driven by questions of the relationship of recreational science to professional science that they have passed over important questions of the relationship of recreational science to recreation.

4. Oldenziel, "Boys and Their Toys" (1997), 63.

5. Bijsterveld, "What Do I Do with My Tape Recorder . . . ?" (2004).

6. O'Connell, "The Fine-Tuning of a Golden Ear" (1992), 5; Perlman, "Golden Ears and Meter Readers" (2004), 789.

7. Clubs, *Movie Makers*, January 1948, 34–35, 37; Wayne Green, "*CQ*'s Survey," *CQ*, December 1957, 34–35; *Amateur Movie Makers*, December 1926, 6; Carl Helmers, "What is *BYTE*?," *Byte*, September 1975, 5.

8. The literature connecting technology to identity includes: Kittler, *Gramophone, Film, Typewriter* (1986); Post, *High Performance* (1994); Pinch and Trocco, *Analog Days* (2002); Gitelman, "Souvenir Foils" (2003); and McCarthy and Wright, *Technology as Experience* (2004). Marshall McLuhan's reference to the "personal" and "psychic" consequences of technology in *Understanding Media* (1964) suggests that "identity" language may in part be an update to an older idea.

9. Turkle, *The Second Self* (1984) and *Life on the Screen* (1995). Through the MIT Initiative on Technology and Self, Turkle is broadening this type of analysis to other technologies.

10. Gitelman also described, without naming them as such, several technical cultures including professional stenographers' "culture of inscription where shorthand made sense and the phonograph made trouble" and the "unconscious culture of letters" formed by those who wrote to Edison, "a folk culture in which individuals asserted their connectedness to the trends and machinations of modern life." Gitelman, *Scripts, Grooves, and Writing Machines* (1999), 63–64, 75.

11. D. Nye, *Electrifying America* (1990); Marvin, *When Old Technologies Were New* (1988).

12. Gitelman, *Scripts, Grooves, and Writing Machines* (1999), 16; Gitelman and Pingree, *New Media, 1740–1915* (2003), xii. Reaching agreement on an identity for a

technology can be thought of in analogy to the way proponents of the social construction of technology describe a technology reaching closure following a period of interpretative flexibility. See Bijker, Hughes, and Pinch, *The Social Construction of Technological Systems* (1987); and Bijker, *Of Bicycles, Bakelites, and Bulbs* (1995).

13. Gelber, *Hobbies* (1999), 35.

14. "Tinkerers," in The Talk of the Town, *The New Yorker*, 31 December 1960, 17–19; *The Best of Groucho* (1955).

15. S. Douglas, "Audio Outlaws" (1992), 53; Barbara Henderson, "Computer Widow," *Kilobaud*, January 1977, 99.

16. News Items, *The Modelmaker*, March 1924, 31; Winter, *The Model Aircraft Handbook* (1942), 248.

17. Jones, *The Encyclopaedia of Early Photography* (1911), v; The Beginners' Section, *Home Photographer and Snapshots*, May 1935, 87–88.

18. Hall, *Home Handicraft For Boys* (1935), vii.

19. For two examples among many, see Wood and Smith, *Prevocational and Industrial Arts* (1919); and LaVoy, *Problems and Projects in Industrial Arts* (1924). Wilber and Neuthardt, *Aeronautics in the Industrial Arts Program* (1942), 2.

20. Our Foreword, *The Modelmaker*, December 1925, 178–179.

21. Frank C. Morton, "Sensitizing Paper at Home," *American Amateur Photographer*, December 1890, 455–457; Hamburg, *Beginning to Fly* (1928), 163.

22. Rodger, "So Few Earnest Workers" (1984), 77; Hamburg, *Beginning to Fly* (1928), 166. For a sociological analysis of the roles and perceptions of "serious leisure," see Stebbins, *Amateurs* (1979); *Amateurs, Professionals, and Serious Leisure* (1992); and *New Directions in the Theory and Research of Serious Leisure* (2001).

23. Zimmermann, *Reel Families* (1995), 113–119; O'Connell, "The Fine-Tuning of a Golden Ear" (1992), 4–5, 10–11, 29–30.

24. Spigel, *Make Room for TV* (1992), 73.

25. R. W. W. [Roy W. Winton], "The Significance of the First Amateur Film Contest," *Movie Makers*, June 1928, 373. Quotations from publications of the Dutch Society of Sound Hunters from the early 1960s in Bijsterveld, "What Do I Do with My Tape Recorder . . . ?" (2004), 625.

26. Perlman, "Golden Ears and Meter Readers" (2004), 802. For the comparison to TRS-80 hobbyists, Perlman builds on Lindsay, "From the Shadows" (2003).

27. Quoted in Levy, *Hackers* (1984), 200.

28. Jones, *The Encyclopaedia of Early Photography* (1911), 22. I was led to this source by Mensel, "Kodakers Lying in Wait" (1991), n. 6.

29. R. W. W. [Roy W. Winton], "Editorial Query," *Amateur Movie Makers*, May 1928, 295; Gelber, *Hobbies* (1999), 53–56.

30. In only a couple cases have I been able to locate data on technical hobbyists' employment: 73% of hams in the 1960s and 57% of audiophiles in the late 1980s and early 1990s earned a living in technical fields. Yet it was commonly believed, and often stated without supporting evidence, that a comparable overlap existed in other technical hobbies as well. Stanford Research Institute, *Amateur Radio* (1966), 31; Perlman, "Golden Ears and Meter Readers" (2004), 789.

31. Our Foreword, *The Modelmaker*, March 1924, 18; Richard E. Byrd, introduction to Hamburg, *Beginning to Fly* (1928), xv; Winter, *The Model Aircraft Handbook* (1942), vi. On boyhood model airplane construction as a guarantee of future aviation progress, see also F. Collins, *The Boys' Book of Model Aeroplanes* (1910).

32. Any nonstandard use—not just the productive pursuits of hobbyists—can alter perceptions about technology. But when entire groups of hobbyists do unusual things with technology, they tend to draw more attention than do single users. For an introduction to the growing literature on the influence of users, see Oudshoorn and Pinch, *How Users Matter* (2003).

33. Quoted in Kittler, *Gramophone, Film, Typewriter* (1986), 81.

34. Developing innovative, leisure uses for technologies permits experimentation with technical identity and values much as the leisure adoption of new forms of popular entertainment, fashion, and language permits experimentation with identity and values generally. On the enactment of culture through leisure, see Peiss, *Cheap Amusements* (1986). Hobbies are just one way among many of coming to grips with new technologies. Two studies commenting on how authors and readers took small steps toward gradually bringing about shifts in technical culture are noteworthy. Laura Otis shows how late nineteenth century fiction writers examined the implications of telegraphy in "The Other End of the Wire" (2001). Joseph Corn and Brian Horrigan point out that popular science magazines of the early twentieth century "interpreted scientific and technological changes for a lay audience." Corn and Horrigan, *Yesterday's Tomorrows* (1984), 6.

35. Mensel, "Kodakers Lying in Wait" (1991), 25.

36. Eastman to W. J. Stillman, 6 July 1888, quoted in Jenkins, "Technology and the Market" (1975), 17.

37. Wilber and Neuthardt, *Aeronautics in the Industrial Arts Program* (1942), 2.

38. Corn, *The Winged Gospel* (1983), 16, 113, 114.

39. Mensel, "Kodakers Lying in Wait" (1991), 29; Keightley, "'Turn it down!' she shrieked" (1996).

40. Bill Gates's letter, dated 3 February 1976, was published in *Computer Notes* (a newsletter for owners of Altair personal computers) and reprinted in many other electronics magazines.

2 The Culture of Ham Radio

1. Carl Dreher and Zeh Bouck, "Our Radio Amateurs," *Harper's Magazine*, October 1941, 535–545, on 535; Jack Cluett, "Listen Here," *Woman's Day*, January 1950, 10–12.

2. Pyle, *ABC's of Ham Radio* (1965), 110; Berens, *Building the Amateur Radio Station* (1959), 115; Fisk, *Ham Notebook* (1973), 4.

3. Paul M. Segal, "The Amateur's Code," in American Radio Relay League, *The Radio Amateur's Handbook* (1969), 6. The identical code was reprinted over many decades, but not always attributed to an author.

4. Maurice J. Hindin, "How Effective Is Amateur Radio Self-Policing?," *CQ*, March 1968, 57–58; Berens, *Building the Amateur Radio Station* (1959), 124; Grammer, *Understanding Amateur Radio* (1976), 289.

5. "It Seems to Us," *QST*, July 1941, 7–8; *The DXer*, October 1948.

6. Hams successfully lobbied state departments of transportation as early as 1950 for the right to use call signs as license plates numbers, decades before the general public was allowed to purchase vanity license plates.

7. Neil D. Friedman, "Amateur Radio Licensing: A Seven-Decade Overview," *QST*, March 1985, 47–48.

8. "Zero Bias," *CQ*, March 1966, 7, 14; Gregory L. Winters, letter to the editor, *CQ*, December 1969, 8, 10; Edward M. Ryan, letter to the editor, *CQ*, February 1970, 7. Two other letters published with Ryan's also spoke out against incentive licensing. "The 1968 *CQ* Reader Survey Report," *CQ*, April 1968, 36–39, 114, 116, on 37.

9. Dannenmaier, *We Were Innocents* (1999), 90.

10. William Ryburn, "In Defense of C.W. [Morse code]," *CQ*, December 1970, 35–37, 81. This esteem for the code continued into the 1990s; see Halprin, *The Code Book* (1993), 7, 11.

11. George Hart, "The CW Language," foreword to Halprin, *The Code Book* (1993), 6.

12. Wayne Green, "*CQ*'s Survey," *CQ*, December 1957, 34–35; "Membership Survey, July 1973, Summary of Responses," *The DXer*, October 1973.

13. Halprin, *The Code Book* (1993).

14. Pyle, *ABC's of Ham Radio* (1965), 10; Carol Witte, letter published in Louisa Dresser, "The YL's Frequency," *CQ*, January 1948, 56, 62.

15. Federal Communications Commission, "In re Application of Myron Henry Premus, W2OY, Lancaster, New York, For Renewal of Amateur Radio Station License and Amateur Radio Operator License," Docket Number 10114, 2 February 1953, copy in "L-3 Early Amateur History—Stories—ARRL" folder, Antique Wireless Association.

16. Howard S. Pyle, "Do You Know the Phillips Code?," *CQ*, January 1967, 55, 100. Sociologists long have recognized specialized language as a boundary marker between social groups. For one description of "self-segregation" through "the use of an occupational slang which readily identifies the man who can use it properly [...] and quickly reveals as an outsider the person who uses it incorrectly or not at all," see Becker, *Outsiders* (1963), 100.

17. Grammer, *Understanding Amateur Radio* (1976), 283; "Zero Bias," *CQ*, April 1966, 6–7. Citizens' Band operators, it is only fair to point out, had their own jargon, including a system called "10-signals," developed by varying a numerical code used for shortening police radio communications. "10–20" functioned the same for CB users and police, and like "QTH" for hams, asking or telling a location. For the police 10-signals not relevant on the Citizens' Band—such as "10–11," used for reporting a "dog case"—CB operators simply changed the meaning, so that "10–11" on the CB became a request to "speak slower." McPherson and Belt, *Easi-Guide to CB Radio for the Family* (1975), 24–25.

18. Don F. E. Fox, "Have You Mumble-itis?," *CQ*, May 1970, 59, 89–90. For the related example of the speech training forced upon the first female telephone switchboard operators, see Martin, *Hello, Central?* (1991), 92–95.

19. Grammer, *Understanding Amateur Radio* (1976), 282.

20. Louis DeLaFleur, "The FCC and the Amateur: Co-Operation Curtails Illegal Operation and Promotes Amateur Prestige," *CQ*, February 1946, 30, 48–50. Joe Craggs to W2BNA, Rochester, New York, 24 June 1954; and Federal Communications Commission, "In re Application of Myron Henry Premus"; copies of both in "L-3 Early Amateur History—Stories—ARRL" folder, Antique Wireless Association.

21. C. Tierney, "The Not-So-Silent SWL [shortwave listener]," *CQ*, September 1953, 17, 46.

22. "Zero Bias," *CQ*, June 1976, 5; *73, Official Publication of the Federation of Radio Clubs of the Southwest*, June 1935, 3.

23. *The DXer*, January 1962. Fred Huntley founded the Anti-Communist Amateur Radio Network to encourage hams to "use their equipment to protect their hobby and their country from conquest on the installment plan." W6RNC [Fred Huntley], letter to editor, *73 Magazine*, December 1961, 68.

24. "Zero Bias," *CQ*, July/August 1970, 5.

25. Wayne Green, "de [from] W2NSD," *CQ*, June 1955, 11; and November 1958, 9, 14, 168.

26. E. M. Wagner, "Need QSO's Be Dull?," *CQ*, November 1967, 78, 114; Grammer, *Understanding Amateur Radio* (1976), 284; DeMaw, *W1FB's Help for New Hams* (1989), 103.

27. American Radio Relay League, *The Radio Amateur's Handbook* (1975), 661. The description of the Rag Chewers Club Award remained largely unchanged since the early 1950s and consistently included the phrases quoted here.

28. See, for example, Carl Perko, "Radio's Rugged Ranks," *CQ*, March 1955, 20–21, 54–56.

29. Ray De Vos, "Pse QSL Tnx [Please QSL; Thanks]," *CQ*, May 1957, 58–59; Otto L. Woolley, "Designing the QSL," *CQ*, January 1950, 16–17. To get an overview of QSL styles, consult the collection of confirmation cards received by one hobbyist and later published in Gregory and Sahre, *Hello World: A Life in Ham Radio* (2003).

30. Weldon Johnson, "QSLs the Photographic Way," *CQ*, September 1957, 42–43, 110.

31. Box 35, "K2AE Ham Letters; QSL cards; photographs, 1958" folder, Broughton Papers.

32. Ingram, *Secrets of Ham Radio DXing* (1981), 47; McCarthy, *CB'ers Guide to Ham Radio* (1979), 65.

33. Louisa B. Sando, "The YL's Frequency," *CQ*, May 1951, 39–40, 68; McCarthy, *CB'ers Guide to Ham Radio* (1979), 191. Entire clubs also visited each other. Carl A. Vidnic, "Ham Clubs Go Visiting," *West Coast Ham Ads*, November 1957, 10.

34. Hertzberg, *So You Want to be a Ham* (1960), 167; *West Coast Ham Ads*, December 1956, 31.

35. The remaining three benefits listed were the monthly newsletter, updates on FCC rule changes, and representation to the national level of ham organizations. "The Rochester Amateur Radio Association (RARA) is the official organized headquarters

for amateur radio in Rochester and vicinity [...]," between pp. 2 and 3 in copy of *The RaRa Rag*, September 1953, bound by Lincoln A. Cundall, Antique Wireless Association.

36. Weiss, *History of QRP in the U.S.* (1987), vi.

37. "It Seems to Us," *QST*, February 1942, 7–13, 45, on 45; *The DXer*, October 1948; *RDXA Bulletin*, December 1950. Dick Talpey, "Forward," [1950?]; and "Minutes of a Special Rochester DX Club Ad Hoc Committee Meeting," 19 September 1979; both loose with *RDXA Bulletins*, Antique Wireless Association.

38. "Zero Bias," *CQ*, March 1950, 9–10; Pyle, *ABC's of Ham Radio* (1965), 112–113.

39. Edwin A. Fensch, "Editing A Club Newspaper," *CQ*, June 1963, 34, 89, 91; Julian Jablin, "'Do-It-Yourself' Club Newspapers," *QST*, March 1958, 54–56; *73, Official Publication of the Federation of Radio Clubs of the Southwest*, June 1935, 3. On the responsibilities of newsletter editors, see Hector E. French, "Roll Your Own," *CQ*, February 1957, 34–36.

40. Carl Perko, "Radio's Rugged Ranks," *CQ*, March 1955, 20–21, 54–56.

41. Employers valued women's manual dexterity and found it financially advantageous to hire women, who worked for lower wages and retired at marriage without collecting pensions. Otis, "The Other End of the Wire" (2001), 194.

42. Unequal dismissal in the wake of the termination of government contracts at the end of the war temporarily interrupted women's employment in electronics. Almost all of the 1,000 workers (one third of the total employees) laid off by Collins in September 1945 were women. Braband, *History of Collins Radio Company* (1983), ii, 44–47, 57–58; Dachis, *Radios By Hallicrafters* (1996), 220.

43. Epstein, "Anti-Communism, Homophobia, and the Construction of Masculinity" (1994); Costigliola, "Unceasing Pressure for Penetration" (1997); and Cuordileone, "Politics in an Age of Anxiety" (2000).

44. S. Douglas, *Inventing American Broadcasting* (1987), 190–194, 203–205.

45. Ingram, *Secrets of Ham Radio DXing* (1981), 9; Orr, *Better Shortwave Reception* (1957), 40; "Zero Bias," *CQ*, September 1968, 5; Tom McMullen, "Focus and Comment," *Ham Radio Horizons*, April 1977, 6.

46. M. Smith, "Selling the Moon" (1983); *RCA Ham Tips*, January–March 1950, 4; *RCA Ham Tips*, Winter 1964–1965, 8; G. D. Hanchett, "The Make-Your-Own Microphone," *RCA Ham Tips*, September 1956, 1–3.

47. *The DXer*, July 1948, 1; G. M. Carrier, "Sued for TVI," *CQ*, October 1957, 50–51. (Contrary to the article title, the author was not actually sued, only threatened with legal action.) Robert M. Ryan, "One Solution," *CQ*, February 1950, 26, 58.

48. Jim Maxwell, conversation with author, Los Gatos, California, 6 June 2000; *The DXer*, June 1961; Ross W. Forbes, letter to the editor, *The DXer*, July 1972; *The DXer*, August 1972.

49. *The DXer*, December 1962 and February 1967; Jerry Hagen, "DX," *CQ*, September 1973, 58–61.

50. *The DXer*, February 1961 and February 1960; Hagen, "DX." Hagen listed the *West Coast DX Bulletin* as one of his sources of information on the event. *The DXer*, February 1974 and April 1974.

51. Fischer, "Changes in Leisure Activities" (1994).

52. Hertzberg, *So You Want to be a Ham* (1960), 168. Also, "Zero Bias," *CQ*, April 1950, 9–10, 69, 72; Michael Colvin, "A Public Relations 'How to' Guide," *Ham Radio Horizons*, August 1979, 42–49.

53. Byron C. Sharpe, "Room with a View: Amateur Radio Links a World-Wide Fraternity," *The Rotarian*, November 1958, 36–38; E. J. Haling, "For a Thrill—Call 'CQ,'" *The Rotarian*, August 1934, 30, 58; Tom Charles, "'CQ-ing' for Goodwill," *The Rotarian*, February 1940, 41–44. *The Rotarian*'s "Hobby Hitching Post" column featured radio amateurs in November 1946 (68–69) and August 1955 (62–63).

54. "Troop 510 B.S.A. Starts Ham Program: Scouting and ham radio join forces to their mutual advantage in this training program," *Radio and Television News*, August 1949, 58, 60; Boy Scouts of America, *Radio* (1947), 11, 10, 23. The reminiscences of hobbyists support the claim that many hams learned about radio through Boy Scouting. One example appears in Weiss, *History of QRP in the U.S.* (1987), 5.

55. "Boy Scouts Need Friends Among Licensed Hams in Tests for Merit Badges," *The RaRa Rag*, January 1960, 3; "Troop 510 B.S.A. Starts Ham Program," *Radio and Television News*; "Zero Bias," *CQ*, March 1950, 9–10; "All City Libraries to Exhibit Ham Stations," *Radiogram*, March 1960, 1, 3. On the related case where the Fisher Body Company promoted its model-building program for boys and young men through the all-male networks of the YMCA and Boy Scouts, see Oldenziel, "Boys and Their Toys" (1997).

56. Weinstein, "Dayton Hamfidential," *CQ*, July 1971, 25.

57. Amelia Black, "The YL's Frequency," *CQ*, April 1946, 44, 48, 50; Eleanor Wilson, "YL News and Views," *QST*, February 1952, 60, 118.

58. "L. I. [Long Island] Woman Trains Others In Army Radio System," *New York World Telegram*, 24 December 1942, clipping in Box 316, Folder 3, Radioana Collection; Karl Detzer, "Biggest Party Line," *Recreation*, March 1947, 627–629; "Friends in Radioland," *Time*, 28 July 1961, 58–59.

59. Amelia Black, "The YL's Frequency," *CQ*, April 1946, 44, 48, 50; Charlene Babb Knadle, "Where Are the Women?," *CQ*, August 1976, 72.

60. Light, "When Computers Were Women" (1999), 461; Oldenziel, *Making Technology Masculine* (1999), 178–179; and Meyerowitz, "Beyond the Feminine Mystique" (1993), 1460.

61. Ellen Marks, "Reply to a YL: 'FB OM, Ur Solid Copy,'" *CQ*, July 1977, 60, 86.

62. "SARA Members June 1962," loose in Box 25, Broughton Papers; *The DXer*, June 1970 and May 1972; "Post-War Hamfestivities," *CQ*, November 1945, 32–33, 43.

63. Carole F. Hoover, "'Club' Your Women," *CQ*, August 1960, 98–99, 101.

64. Jim Maxwell, conversation with author, Los Gatos, California, 6 June 2000; *The DXer*, February 1964 and June 1973.

65. M. Smith, "Selling the Moon" (1983); *Dr. Strangelove* (1964).

3 Equipping Productive Consumers

1. Adorno, "On the Fetish-Character in Music and the Regression of Listening" (1938), 293. My thanks to Mark Lewis for referring me to this essay.

2. On the mixture of production and consumption and of work and recreation in hobbies generally, see Gelber, *Hobbies* (1999).

3. Orr, *Better Shortwave Reception* (1957), 82; Helfrick, *Amateur Radio Equipment Fundamentals* (1982), 190; Dezettel, *Realistic Guide to Electronic Kit Building* (1973), 13. "Realistic" was the name of Radio Shack's proprietary brand.

4. de Henseler, *Hallicrafters Story* (1991), 277; DeSoto, *Two Hundred Meters and Down* (1936), 123; Orr, *Better Shortwave Reception* (1957), 40.

5. Berens, *Building the Amateur Radio Station* (1959), v. Although the title suggests otherwise, *Building the Amateur Radio Station* provided a general introduction to the hobby and covered far more than construction. One of few changes in the second edition of this handbook was the complete removal of the chapter on "Commercially Available Ham Equipment," as if the authors were trying to blind hams to the availability of readymade equipment and thereby force them into the building experience. Berens and Berens, *Building the Amateur Radio Station* (1965).

6. "In Our Opinion," *CQ*, June 1960, 6; "$1000 Cash Prize 'Home Brew' Contest," *CQ*, March 1950, 17. This considerable prize money was to be divided between ten winners. Wayne Green, "de [from] W2NSD," *CQ*, November 1958, 9, 14, 168.

7. Ingram, *44 Electronics Projects* (1981), 39; Weiss, *History of QRP in the U.S.* (1987), 9; Findlay, *Electronic Experimenter's Manual* (1959), 48; Ed Noll, "Circuits and Tech-

niques," *Ham Radio Magazine,* July 1971, 58–62. Wood presumably stood out among radio shack electronics as particularly amateurish because of the deep-seated associations of wood with old-fashioned craft and of metal with modernity and technical progress. On the design preference for metal over wood for symbolic reasons, see Schatzberg, *Wings of Wood, Wings of Metal* (1999).

8. These terms for the first radio users appear in many sources, including Duston, *Radio Theory Simplified* (1926), 170.

9. A thorough description of the practices involved in the hobby of radio before and at the beginning of broadcasting can be found in S. Douglas, *Inventing American Broadcasting* (1987). The copy of *The EKKO Broadcasting Station Stamp Album* (1924) at the American Radio Relay League Library includes an impressive collection of reception stamps. Photographs of a listener's log book from the mid 1920s and stations' confirmation cards of the late 1930s are published in Jaker, Sulek, and Kanze, *The Airwaves of New York* (1998), 4, 20.

10. The cost-saving strategy of building radios is explained with regard to industrial workers' participation in the hobby of radio in Cohen, *Making a New Deal* (1990), 129, 132–143. Duston, *Radio Construction For the Amateur* (1924), 20–21.

11. Godley, *Getting Acquainted with Radio Receivers* (1923). Although its title is general, the book discusses only the particulars of the Paragon Receiver, and the cover of the copy in the American Radio Relay League Library is stamped: "This booklet is a part of the Paragon Receiver, with which it is packed, and must be delivered to the purchaser of the receiver."

12. O. H. Caldwell, "The Radio Market," in *Radio and Its Future* (1930), ed. Codel, 203–210, on 203.

13. *Radio Broadcast,* November 1925, 67. Protective policy quoted in Carlat, "A Cleanser for the Mind" (1998), 122. Carlat's essay is especially strong on how considerations of gender influenced manufacturers' simplification of radios and stifling of tinkering. For an analysis of radio design aesthetics, see Nickles, "Object Lessons" (1999).

14. Jome, *Economics of the Radio Industry* (1925), 69, 82; S. Douglas, *Inventing American Broadcasting* (1987), 293–295, 299–303; David Sarnoff, "Art and Industry," in *Radio and Its Future* (1930), ed. Codel, 185–195, on 187.

15. Duston, *Radio Theory Simplified* (1926), 170; advertisement reproduced in Cones and Bryant, *Zenith Radio* (1997), 13.

16. Braband, *History of Collins Radio Company* (1983); de Henseler, *Hallicrafters Story* (1991), 15; Dachis, *Radios By Hallicrafters* (1996), 9. William Halligan also was

instrumental in helping Theodore Deutschmann establish the first Radio Shack store in Boston in 1921. Farman, *Tandy's Money Machine* (1992), 114.

17. To get a sense of the role of amateur radio sales within large, diversified radio-electronics companies, I consulted General Electric Co., *Annual Report* (1945–1960); and Radio Corp. of America, *Annual Report* (1927–1960).

18. E. E. Williams, "Commercial Aspects: Emergency and Miscellaneous Communication Equipment and Apparatus," in *Proceedings of General Electric Radio Specialists' Meeting* (1940).

19. E. H. Fritschel, "The Radio Transmitting Tube Market and our Position In It," in *Proceedings of General Electric Radio Specialists' Meeting* (1940).

20. Braband, *History of Collins Radio Company* (1983), 42; de Henseler, *Hallicrafters Story* (1991), 23; "Sets for the Hams," *Business Week*, 24 November 1945, 41–42.

21. Hallicrafters Co., *Annual Report* (1950), 5; "Sets for the Hams," *Business Week*; U.S. Federal Communications Commission, *Annual Report* (1947), 55; Lindemann, "Marketing Communications Receivers" (1947), 7.

22. Bill Orr, "The War Surplus Story," *CQ*, Wrap-Up 1977 (issue 13), 50–54. Articles on surplus equipment were indexed in *CQ*, January 1956.

23. "Report Significant Gains By Parts Distributors In Industrial Sales," *Jobber News and Electronic Wholesaling*, August 1957, 1. (The name "jobber" in this context refers to an electronics parts dealer.) "Distributor Survey Report," *CQ*, November 1959, 132–141, 146, on 132; *Schenectady Amateur Radio Association Newsletter*, January 1956, 4. H. W. Barber, Sales Promotion Section, GE Electronics Department, Tube Division, Schenectady, "To G-E Transmitting Tube Distributors" (memorandum sent with the first issue of *Ham News*), 16 May 1946, Box 15, Broughton Papers.

24. Hallicrafters Co., *Annual Report* (1950), 3; "Distributor Survey Report," *CQ*, 132.

25. Braband, *History of Collins Radio Company* (1983), 102, 72; Wayne Green, "*CQ*'s Survey," *CQ*, December 1957, 34–35; "1968 *CQ* Reader Survey Report," *CQ*, April 1968, 36–39, 114, 116; Stanford Research Institute, *Amateur Radio* (1966), 31; "Amateur Radio Operators In Company Wish to be Listed in G. E. Directory of 'Hams,'" *G. E. Co. River Works News*, 12 June 1936, clipping in Series 136, Box 385, Folder 2, Radioana Collection; Heath Co. advertisement, *CQ*, September 1958, 4–8; Allied Radio Corp., *Allied Radio*, catalog no. 220 (1963), 28. The power of habit and emotion as market forces is difficult to document. For a study of how engineers and managers with a personal interest in small radios persuaded Raytheon to pursue the design of such a product, see Schiffer, "Cultural Imperatives and Product Development" (1993).

26. "Zero Bias," *CQ*, January 1967, 5; and September 1970, 5.

27. Hallicrafters Co., *Hallicrafters Radio,* catalog no. 36 (1945).

28. Morton B. Kahn, "75 Watts in Full Dress," *CQ,* September 1946, 24, 57–58; Collins Radio Co., "The Collins 75A-1 Amateur Receiver" (1948), 3; Allied Radio Corp., *Allied Radio,* catalog no. 124 (1951), 137.

29. Alexander MacLean, "Collins R390A modifications: Several simple modifications for the R-390A which can considerably improve performance," *Ham Radio Magazine,* November 1975, 66–67; William I. Orr, "The Collins 310B—1953 Version," *CQ,* June 1953, 13–18, on 13; J. Miller, *The Pocket Guide to Collins Amateur Radio Equipment* (1995), 89; Mims, *Integrated Circuit Projects,* (1973), 1:3.

30. Radio Corp. of America, *Common Words in Radio, Television and Electronics* (1947); Radio Corp. of America, *What's the Right Word?* (1952).

31. Wayne Green, "*CQ*'s Survey," *CQ,* December 1957, 34–35; James N. Brink, "Building The Heathkit 'Comanche' Mobile Receiver," *CQ,* October 1959, 52–53, 109–111; Hertzberg, *So You Want to be a Ham* (1960), 187; R. Brown, *104 Easy Projects For the Electronics Gadgeteer* (1970), 11; Levey, *Radio Receiver Laboratory Manual* (1956), 105. Levey taught at the Long Island Agricultural and Technical Institute and intended the *Radio Receiver Laboratory Manual* for use in radio engineering courses.

32. Hammarlund Manufacturing Co., "HQ-120-X" [1940] and "Hammarlund HQ-120-X Receiver" [1940].

33. Allied Radio Corp., *Allied Radio* (1939), 124; Mims, *Integrated Circuit Projects* (1973), 2:96, 1:7.

34. "'Ham Tips' Announces Contest for RCA Tube Equipped Rig," *RCA Ham Tips,* July 1946, 1; H. W. Barber, Sales Promotion Section, GE Electronics Department, Tube Division, Schenectady, "To G-E Transmitting Tube Distributors" (memorandum sent with the first issue of *Ham News*), 16 May, Box 15, Broughton Papers; *The DXer,* August 1962.

35. Pollack, *Transistor Theory and Circuits Made Simple* (1958).

36. Hydro-Aire, *The Transistor and You* (1955), 3, 4.

37. Capstone Electronics Corp. Technical Staff, *How to Use Bargain Transistors* (1967), 61. Guidance on the interchangeability of transistors also was available in reference books like *Transistor Specifications and Substitution Handbook* (1966).

38. Steven Gelber's chapter examining the place of kits within hobby culture, "Kits: Assembly as Craft," reaches some conclusions about kits overall that ring true for ham radio and electronics kits. For example, the directed work of kits erased freedom and diminished the skills required. Other generalizations about kits, such as that they played a role in bringing gender neutrality to hobbies, do not apply to the case of electronics kits. Gelber, *Hobbies* (1999), 255–267.

39. "Zero Bias," *CQ*, February 1950, 9. Wayne Green, "de [from] W2NSD," *CQ*, April 1956, 11–12; and November 1958, 9, 14, 168.

40. Beason, *How to Build Electronics Kits* (1965), 2; Allied Radio Corp. advertisement, in Shuart, *Radio Amateur Course* (1937), 71.

41. Allied Radio Corp., *Allied Radio* (1935), 83; Hammarlund Manufacturing Co., "Transmitter Foundation Units" (1940).

42. Gelber, *Hobbies* (1999), 261–267.

43. Heath Co. advertisement, *CQ*, September 1958, 4–8. The broad outlines of Heath's history are widely reported. This summary particularly draws on Chuck Penson, "Heathkit's 50th: The Green Turns to Gold," *Electric Radio*, January 1997; and Perdue, *Heath Nostalgia* (1992). Wayne Green, "*CQ*'s Survey," *CQ*, December 1957, 34–35.

44. Allied Radio Corp., *Allied Radio*, catalog no. 220 (1963), 27.

45. These statements appeared repeatedly in Heathkit manuals, typically in the warranty or as part of the "Helpful Kit Building Information."

46. "Zero Bias," *CQ*, February 1950, 9; and March 1950, 9–10. Dezettel, *Realistic Guide to Electronic Kit Building* (1973), 86; Beason, *How to Build Electronics Kits* (1965), 53; Frank Beacham, "Bidding a Fond Farewell to The Do-It-Yourself Heathkits," *Radio World*, 10 June 1992, 15.

47. "1968 *CQ* Reader Survey Report," *CQ*, April 1968, 36–39, 114, 116, on 39.

4 Amateurs on the Job

1. "Zero Bias," *CQ*, January 1954, 11.

2. F. Barrows Colton, "Your New World of Tomorrow," *National Geographic*, October 1945, 385–410, on 399; *The Standard Periodical Directory* (1964), 151; Carl Dreher, "Electronics Promises New Miracles in Industry," *Popular Science Monthly*, September 1944, 128–133, 208–209.

3. "Electronics: A Lever on Industry," *Fortune*, July 1943, 133–135, 198, 200, 202, 205, on 133, 198; "Electronics Era," *Business Week*, 29 July 1944, 24, 26, 29–30; Carl Dreher, "What's Ahead in Electronics?," *Popular Science Monthly*, July 1944, 68–73, 214, 216, 218, 221, on 68; Dreher, "Electronics Promises New Miracles in Industry," 128. The other articles in Dreher's series were "What About Postwar Television?," *Popular Science Monthly*, August 1944, 78–83, 228, 232, 236; and "Electronics Brings Magic New Aids to Better Living," October 1944, 136–140, 217–218. Similar articles addressing the implications of electronics for business included "Electronic Era: Rapidly growing industry agrees that it must strive for practical utility, not glamour, in

postwar developments," *Business Week*, 14 October 1944, 65–66; and Keith Henney, "Electronics In Tomorrow's Industry," *Scientific American*, November 1944, 208–210.

4. "Electronics: A Lever on Industry," *Fortune*, 133; John Sasso, "What's All This About Electronics?," *House Beautiful*, May 1943, 31, 123, 127; Walter Adams, "Mystery Weapon Today, Your Servant Tomorrow," *Better Homes and Gardens*, August 1943, 20–21, 64–67.

5. "Electronic Gadgeteering May Replace 'Hams'' Interest in Communication," *Scientific American*, July 1942, 28.

6. "Electronics Era," *Business Week*, 29 July 1944, 24, 26, 29–30; William E. Taylor, "Electronically Yours, G. I. Joe," *Radio-Electronic Engineering*, wholly contained within *Radio News*, February 1945, 36–37, 128; U.S. Dept. of Labor, Bureau of Labor Statistics, Bulletin No. 940 (1948), 66. Veterans also took their skills into other industries and hobbies. War-trained pilots, for example, entered commercial aviation and greatly increased the number of participants in recreational flying. F. Barrows Colton, "Your New World of Tomorrow," *National Geographic*, October 1945, 385–410.

7. These figures disregarded the federal government, universities, and nonprofit research centers as employers, which together kept an additional 55,000 electronics workers on the payroll in 1960. U.S. Dept. of Labor, Bureau of Labor Statistics, Bulletin No. 1363 (1963), 1.

8. Kneitel, *Jobs and Careers in Electronics* (1960), 1; B. Richard, "Electronic Servicemen and Technicians: One of the Fastest-Growing Occupations in the U.S.," *Popular Mechanics*, July 1960, 32–33; "When the Brains Can't Get Work," *Business Week*, 13 February 1971, 90–92, 94; Jim Calder, "Is Electronics Still a Good Career?," *Mechanix Illustrated*, December 1974, 34–35, 99; Art Salsberg, "Where Are You Going in Electronics?," *Popular Electronics*, May 1980, 4.

9. Findlay, *Electronic Experimenter's Manual* (1959), 1; U.S. Dept. of Labor, Bureau of Labor Statistics, Bulletin No. 1072 (1952), 22; Findlay, *Electronic Experimenter's Manual* (1959), 2; Hertzberg, *So You Want to be a Ham* (1960), 3, 177, 182. Of particular note among the endless stream of suggestions that ham radio could lead to an electronics career is the series of articles: Howard S. Pyle, "Electronics Careers: Is There One in Your Future," *CQ*, September 1967, 22–26, 114; October 1967, 33–37; and November 1967, 65–70. Similar remarks appeared in general electronics hobby publications such as: Dezettel, *Realistic Guide to Electronic Kit Building* (1973), 13; and Calder, "Is Electronics Still a Good Career."

10. Membership list, including employment information, stored with *Rochester DX Association Bulletins*, Antique Wireless Association; *The RaRa Rag*, September 1949, 4; McGirr, *Suburban Warriors* (2001), 27.

11. The 1968 survey reported that 42.7% of readers were employed in the electronics industry. "Zero Bias," *CQ*, October 1963, 5; Wayne Green, "*CQ*'s Survey," *CQ*, December 1957, 34–35; "1968 *CQ* Reader Survey Report," *CQ*, April 1968, 36–39, 114, 116, on 37; Stanford Research Institute, *Amateur Radio* (1966), 31.

12. Norman Eisenberg, "Opportunities in Electronics," in *Jobs and Careers in Electronics* (1960), ed. Kneitel, 6–14, on 8; Carolyn Cummings Perrucci, "Engineering and the Class Structure," in *Engineers and the Social System* (1969), ed. Perrucci and Gerstl, 279–310, on 283; Robert L. Eichhorn, "The Student Engineer," in *Engineers and the Social System*, 141–159, on 144.

13. Green, "*CQ*'s Survey"; U.S. Dept. of Commerce, Bureau of the Census, *Historical Statistics of the United States* (1975), 292; Stanford Research Institute, *Amateur Radio* (1966), 32, 83.

14. Jean Kenton, "What You Should Study in School," in *Jobs and Careers in Electronics* (1960), ed. Kneitel, 92–95.

15. Hertzberg, *So You Want to be a Ham* (1960), 184, 182; Ingram, *44 Electronics Projects* (1981), 28.

16. James N. Whitaker, "Hams in Industry," *CQ*, September 1953, 38–39; Hughes Aircraft Co. Research and Development Laboratories advertisement, *CQ*, January 1955, 53. A call for hams in the fields of engineering, programming, and mathematics appeared in *The DXer*, August 1966, 3.

17. Braband, *History of Collins Radio Company* (1983), 41; Collins Radio Co., "The Collins 30K-1 amateur transmitter" (1947), 2.

18. "Amateur Radio Operators In Company Wish to be Listed in G. E. Directory of 'Hams,'" *G. E. Co. River Works News*, 12 June 1936, clipping in Box 385, Folder 2, Radioana Collection; *RDXA Bulletin*, 9 September 1966, 1; Neil A. Martin, "The Hams Are Still in Business," *Dun's Review*, April 1972, 62–65; Bill Welsh, "An Active Amateur Radio Club," *Ham Radio Horizons*, February 1979, 24–30; Jim Maxwell, conversation with author, Los Gatos, California, 9 June 2000. There have been other instances of industry recruiting employees by supporting recreational activities. Ruth Oldenziel documented how a construction contest for boys run by the Fisher Body Company, a supplier of auto bodies to General Motors, served in part "to groom them as technical men ready to take their places as managers or engineers in GM's corporate world." Oldenziel, "Boys and Their Toys" (1997), 65.

19. "About the Emergency Truck," *The RaRa Rag*, May 1957, 7–8; *The RaRa Rag*, February 1950, 2.

20. "There is no easier way to embark on a career in electronics and allied fields," he continued, "than Amateur Radio." Tompkins, *SOS at Midnight* (1957), 223.

21. Cleveland Institute of Electronics advertisement, *CQ*, October 1967, 32.

22. M. Smith, "Selling the Moon" (1983), 179, 177.

23. *The DXer*, October 1960; *The DXer*, June 1962; *The DXer*, January 1963.

24. "Zero Bias," *CQ*, August 1965, 7; and July 1969, 5.

25. To name one example, the Radio Club of America, a New York City club begun as The Junior Wireless Club in 1909, patterned itself on a scientific society. Radio Club of America, *Yearbook* (1930), 3, 4. On professionalization as an aspect of the Progressive era, see Wiebe, *The Search for Order* (1967).

26. My analysis of the tension and "splitting" in hams' identification was inspired by Bhabha, "Interrogating Identity" (1994).

27. Mills, *White Collar* (1951); Whyte, *The Organization Man* (1956). On hobbies as idealized work, see Gelber, *Hobbies* (1999), 33–34.

28. Fisk, *Ham Notebook* (1973), 4; Ingram, *44 Electronics Projects* (1981), 28.

29. Oldenziel, *Making Technology Masculine* (1999), 53.

30. Hallowell Bowser, "Strength on the Bench," *Saturday Review*, 25 February 1961, 32.

31. Ferguson, *Engineering and the Mind's Eye* (1992), 199–200, n. 32, including quotation from M. Rychener, ed., *Expert Systems for Engineering Design* (Boston: Academic Press Inc., 1988), 22–23. Nostalgia for production predating the assembly line often includes reference to workers' integrated facility with things and ideas. See, for example, Harper, *Working Knowledge* (1987).

32. Karl B. Keller, "Protect that Invention: Basic Information About the U.S. Patent System," *QST*, January 1962, 63–65; Robert G. Slick, "How to Profit from Your Electronic Invention," *CQ*, March 1953, 24–26; Paul M. Segal, "The Amateur's Code," in American Radio Relay League, *The Radio Amateur's Handbook* (1969), 6; *The RaRa Rag*, February 1949, 1.

33. *Schenectady Amateur Radio Association Newsletter*, February 1949; A. David Middelton, "Inside the Shack and Workshop," *CQ*, February 1947, 36–37; *The RaRa Rag*, September 1949, 10, 11.

34. For one example among many, see R. D. Valentine and Athan Cosmas, "Another Method of Converting the ART/13," *CQ*, February 1947, 33–35, 66–70; it was a response to Paul L. Rafford Jr., "Converting the ART/13 Transmitter," *CQ*, November 1946, 13–17, 63.

35. Jack Cluett, "Listen Here," *Woman's Day*, January 1950, 10–12.

36. The five letters between Andy Shafer (carbon copies) and Arthur E. Ericson (originals) on the coherer receiver, dated between 17 December 1970 and 13 January 1971 are preserved in the "L-3 Early Amateur History—Stories—ARRL" folder, Antique Wireless Association. Other examples of ham correspondence that include technical drawings are located within the "K2AE Ham Letters; QSL cards; photographs, 1958" folder, Box 35, Broughton Papers.

37. Stanley S. Robin, "The Female in Engineering," in *Engineers and the Social System* (1969), ed. Perrucci and Gerstl, 203–218, on 210; Spigel, *Make Room for TV* (1992), 73.

5 Hobby Radio Embattled

1. Paul W. Kearney, "Calling All Hams!," *This Week*, 16 February 1941; and Will H. Connelly, "It's a Radio War," *Liberty*, 30 December 1944; both clippings in Box 298, Folder 3, Radioana Collection. On the strategic role of communications technology, see Headrick, *The Invisible Weapon* (1991).

2. Kearney, "Calling All Hams"; DeSoto, *Two Hundred Meters and Down* (1936), 52; "Sets for the Hams," *Business Week*, 24 November 1945, 41–42; Stuit, *Personnel Research and Test Development* (1947), 118; E. L. Battey, "Radio as a Hobby in the Navy," *CQ*, November 1951, 27; Snyder, *Air Force Communications Command* (1991), 88; "Naval Reserve Offers Inducement to Hams," *Radiogram*, February 1964, 1; Hertzberg, "The Ham in Military Service," in *So You Want to be a Ham* (1960), 171–176.

3. "It Seems to Us," *QST*, February 1960, 9–10; *The DXer*, May 1966, 1–5; American Radio Relay League, "Year End Membership Totals from 1920–1995," internal record provided to author; Bob Wetherald, "Splatter," *The RaRa Rag*, September 1953, 3.

4. "Clip it out!," *QST*, February 1941, 24; "It Seems to Us," *QST*, February 1960, 9–10; K. B. Warner, ARRL Secretary, 3 November 1922, memorandum "To all Directors, Affiliated Clubs, Division Managers, District Superintendents, and City Managers," "L-3 Early Amateur History—Stories—ARRL" folder, Antique Wireless Association. Other ham radio magazines provided similar public relations guides such as the speech "Amateur Radio—A Public Service" reprinted in *CQ* with the editor's encouragement that hams "take advantage of its excellent phraseology." "A Public Relations 'How to' Guide" in *Ham Radio Horizons* told readers not to allow the ARRL's flashy, mass media commercials—Dick Van Dyke, Bob Hope, Lorne Greene, and Captain and Tennille taped Public Service Announcements for the ARRL in the late 1970s—to make them complacent because "the real work of public relations belongs to local Amateurs and clubs." "Zero Bias," *CQ*, April 1950, 9–10, 69, 72; Michael Colvin, "A Public Relations 'How to' Guide," *Ham Radio Horizons*, August 1979, 42–49, on 42.

5. Allied Radio Corp., *Allied Radio*, Spring/Summer catalog (1941), 137; "A Call to Action," *Amateur Radio Defense*, November 1940, 11–12.

6. "A Call to Action," *Amateur Radio Defense*; Arthur H. Halloran, "He Also Serves," *Amateur Radio Defense*, November 1940, 2; L. R. Huber, "Chats with the Editors," *Amateur Radio Defense*, January 1941, 4.

7. "What the League is Doing," *QST*, August 1940, 24–26; "Operating News," *QST*, January 1941, 44–45; "It Seems to Us," *QST*, June 1941, 7–8.

8. "What the League is Doing," *QST*, July 1940, 22–29, on 23; K. B. W. [Kenneth B. Warner], "F.C.C. Orders and Interpretations: Notes for Your Guidance and Assistance," *QST*, August 1940, 17–18.

9. Box 25, "Proof of USA Citizenship, Birth Ctfcs [Certificates], and XYL" folder, William Broughton Papers. Henry Broughton was William Broughton's father.

10. "Operating News," *QST*, June 1941, 63–64; "What the League is Doing," *QST*, December 1940, 34–35; "Help Wanted in Monitoring," *QST*, September 1940, 23; "It Seems to Us," QST, August 1941, 7–8.

11. "Amateurs Pretend to Be Nazi Spies and it's Not Funny to Uncle Sam," *Variety*, 25 June 1941, clipping in Box 385, Folder 2, Radioana Collection; "Radio Spies Are Trapped by Direction Finders in Prowling Motor Cars," *Popular Science Monthly*, May 1941, 120; "Illegal Transmitter," *Time*, 5 January 1942, 49–50; Carl Dreher and Zeh Bouck, "Our Radio Amateurs," *Harper's Magazine*, October 1941, 535–545, on 544.

12. "What the League is Doing," *QST*, January 1941, 18–19.

13. Hart, *Manual for the War Emergency Radio Service* (1944), 38, 20.

14. U.S. Federal Communications Commission, "Part 15: Rules Governing All Radio Stations in the War Emergency Radio Service," reprinted in "The War Emergency Radio Service," *QST*, July 1942, 11–15; "It Seems to Us," *QST*, July 1942, 9–10.

15. Matthews, *The Specter of Sabotage* (1941), 105; New Jersey Defense Council, *The Control System of the Civil Protection Services* (1942), 15; Minnesota Office of Civilian Defense, *Victory Aide Handbook* [1942 or 1943], 21.

16. "What the League is Doing," *QST*, January 1941, 18–19; "A-ARS Activities," *QST*, November 1941, 51; "It Seems to Us," *QST*, November 1941, 7–9, 62.

17. George Barrett, "Radio Hams in U.S. Discuss Girls, So Shelling of Seoul Is Held Up," *New York Times*, 9 February 1951, late city edition.

18. Heathkit/Zenith Educational Systems, *Amateur Radio General License Course* (1979), 1:54; Nancy Anderson, "Hobbies," *CQ*, April 1956, 87–88; MacDonald,

Television and the Red Menace (1985), 103, 196; Charles J. Schauers, "Electronic Bugging and the Ham," *CQ*, December 1966, 34–36.

19. Allan Keller, "The Spying System for U-Boats Worked Perfectly Until a Radio 'Ham' Tuned in on the Plot," *New York World Telegram*, 2 September 1936, clipping in Box 385, Folder 2, Radioana Collection; Paul G. Watson, "Historical Notes Covering the Early Development of the Electron Tube and Radio Apparatus," vol. 2 (bound manuscript, 1 October 1963), 167, Box 1, Watson Collection; Paul W. Kearney, "Calling All Hams!," *This Week*, 16 February 1941, clipping in Box 298, Folder 3, Radioana Collection. Contemporary hams still retain a reputation for scanning the airwaves and turning in wrongdoers. Dave Eggers recounted his anxiety when scattering his mother's ashes without permission that "someone's going to come here tomorrow and find them and then report it and connect it to me, because the funeral home guy, Chad, with his ham radio, will be listening to a police frequency." Eggers, *A Heartbreaking Work of Staggering Genius* (2000), 341.

20. "Hams Across the Iron," *Time*, 28 February 1949, 62; Louis DeLaFleur, "The FCC and the Amateur: Co-Operation Curtails Illegal Operation and Promotes Amateur Prestige," *CQ*, February 1946, 30, 48–50.

21. Ruth E. Johnson, "Wife's Eye View," *CQ*, July 1946, 18, 60–61; Orr, *Better Shortwave Reception* (1957), 128.

22. "Alleged Red Spy Seized in Hidden Radio Station Raid," *Los Angeles Times*, 18 August 1950, 9 a.m. final edition; "Naval Radar Expert Seized as Russ Spy," *Los Angeles Times*, 19 August 1950, 9 a.m. final edition. "N.Y. Engineer Is 8th Russian Spy Ring Suspect," byline 18 August 1950; and "Spy Radio: Code Books Seized in Mexico Raid," byline 19 August 1950, unidentified clippings in the collection of Jim Maxwell, now housed at the California Historical Radio Society. Sobell, *On Doing Time* (1974), 17. Morton Sobell was sentenced to thirty years in prison for conspiracy to commit espionage, the same charge for which the Rosenbergs were executed.

23. "Zero Bias," *CQ*, January 1954, 11; and February 1954, 11, 50. Joseph McCarthy quoted in "Zero Bias," *CQ*, May 1954, 7, 84; this editorial went on to criticize the ARRL for abstaining from political matters to the point that the League had not told its members of these accusations made against hobbyists. U.S. Federal Communications Commission, *Annual Report* (1954), 29; and (1955), 28. The 1956 FCC *Annual Report* mentioned that the proposal continued to await a decision, but the matter was dropped entirely by the 1957 report.

24. Weiss, *History of QRP in the U.S.* (1987), 4; Theodore M. Hannah, "Amateur Radio, Russian Style," *QST*, November 1958, 61–64, 182, 184, on 61; Theodore M. Hannah, "Russian Amateur Radio—1962 Style," *QST*, August 1962, 80–81; Editorial, *The DXer*, May 1964, 1.

25. Letter to editor from W6RNC [Fred Huntley] and reply from W2NSD [Wayne Green], *73 Magazine*, December 1961, 68–69. The Northern California DX Club, among other groups, endorsed Green's position. Editorial, *The DXer*, January 1962, 1.

26. "QUA," *CQ*, January 1945, 5. One call for nominations for the Edison Award appeared in a General Electric advertisement, *QST*, September 1955, 1.

27. Edward P. Tilton, "A Civil Defense Portable: A Four-Pound Station for Use Wherever Man Can Go on Foot," *QST*, May 1951, 35–38, 118, 120, on 35.

28. Editorial, *QST*, June 1948, 9–10; U.S. Office of Civil Defense Planning, the Hopley Report (1948), 108. For an overview of civil defense in the United States, see Boyer, "The Reassuring Message of Civil Defense," in *By the Bomb's Early Light* (1985), 319–333. On the movement against civil defense, see Garrison, "Our Skirts Gave them Courage" (1994), 201–226.

29. U.S. Office of Civil Defense Planning, the Hopley Report (1948), 117, 298, Chart 8.

30. Editorial, *QST*, December 1950, 9–12.

31. Prentiss, *Civil Defense in Modern War* (1951), 371; U.S. Office of Civil Defense Planning, the Hopley Report (1948), 108.

32. New York State Civil Defense Commission, *Radio Officer's Guide* (1961), 7; U.S. Federal Civil Defense Administration, *Annual Report* (1955), 99; U.S. Office of Civil and Defense Mobilization, *Annual Statistical Report* (1961), 85–86.

33. "The Radio Amateur Civil Emergency Service: Part II—The Communications Plan, Station and Operator Authorizations," *QST*, April 1953, 59–61, 140; U.S. Office of Civil and Defense Mobilization, *Annual Statistical Report* (1961), 85–86.

34. Editorial, *QST*, February 1952, 9; "The Radio Amateur Civil Emergency Service: Part I—What It Is and What It Isn't," *QST*, March 1953, 50–52, 116; "The Radio Amateur Civil Emergency Service: Part II," *QST*; "The Radio Amateur Civil Emergency Service: Part III—Funds and Frequencies," *QST*, May 1953, 60–62, 134, 136.

35. Francis Gary, "Amateur Radio in Detroit Civil Defense: Motor City Amateurs a Vital Link in CD Communications," *QST*, September 1951, 52–53, 110; M. P. Rehm, "A Civil Defense Club Project: Tri-County Radio Association Program Provides Emergency Stations and Promotes V.H.F. Activity in Northern New Jersey," *QST*, October 1951, 15–17, 120, 122; Robert G. Seymour, "C.D. Committee Report," *QST*, July 1953, 60–61; "Operating News," *QST*, February 1952, 62–63. There were four *QST* articles about the construction of equipment for civil defense in 1951 and seven more in 1952–1953.

36. "Fallout Shelters," *Life*, 15 September 1961, 95–108, on 95, 96; Grammer, *Understanding Amateur Radio* (1976), 287.

37. "NPA [National Production Authority] Establishes Priority Aid for Amateur Services," *Radiogram*, December 1951, 1, 4.

38. Data from articles in *QST* by George Hart: "Amateurs in Operation Alert 1955," September 1955; "OPAL 1956," November 1956; "Operation Alert 1957," November 1957; "Operation Alert, 1958: A Symposium of Amateur Participation," October 1958; "Operation Alert 1960: Working Together in Civil Defense," October 1960, which includes data for 1959; and "Operation Alert, 1961: Including an Analysis of Amateur Participation in the Conelrad Alert," August 1961.

39. "Announcing 1960 Simulated Emergency Test, October 8–9, 1960," *QST*, October 1960, 49; George Hart, "Simulated Emergency Test—1961," *QST*, April 1962, 21–25.

40. Cooper and Cone, "Defending the Delaware Home Front" (1995–1996), 249; and R. Miller, "The War that Never Came" (1991), 3.

41. Ballard, "Preparing for the Unthinkable War" (1996); McIlroy, "No Interest, No Time, No Money" (1997); and Winkler, "A 40-year history of civil defense" (1984).

42. "REN [Rochester Emergency Net] News," *The RaRa Rag*, May 1951, 6.

43. R. A. Harris, "How to Gain the Goodwill of the Public," *73, Official Publication of the Federation of Radio Clubs of the Southwest*, June 1935, 7–8. Compare this to a strikingly similar statement about television interference: "neighbors who read about the Christmas message Joe Ham handled for another neighbor will be slower to raise Cain if they should hear his voice on their TV set." "It Seems to Us," *QST*, February 1960, 9–10.

44. "Zero Bias," *CQ*, February 1948, 13; George Hart, "Simulated Emergency Test—1952," *QST*, April 1953, 62–63. Emergency communication drills also ranked high on the list of topics that would interest a broad readership in Charles H. Curran, "Publicity for Your Club—Free!," *CQ*, March 1958, 35, 108.

45. "Announcing 1961 Simulated Emergency Test, Oct 7–8, 1961," *QST*, October 1961, 45; "Zero Bias," *CQ*, September 1964, 7, 90.

46. U.S. Office of Civil Defense, *1964 Annual Statistical Report* (1965), 40. Other sources that cited ham radio as having practical application for backup communication only as part of natural disaster relief include George S. Gibbs, "Maj. Gen. Gibbs Stresses Importance of Amateurs Affiliated With the Army," *New York Herald Tribune*, 29 December 1929, clipping in Box 299, Folder 1; and Paul W. Kearney, "Calling All Hams!," *This Week*, 16 February 1941, clipping in Box 298, Folder 3, both in the Radioana Collection. It is difficult to evaluate the overall importance of ham radio

to post-disaster communication. Hams have exaggerated their role at times, and appreciative relief personnel have been reluctant to rein in hobbyists' claims. One indication of the limited need for and efficacy of two-way radio comes in the transcript of a telephone call placed to a civil defense unit following a hurricane: the ham informs the caller, who has requested that a message be relayed by radio to a family member, that she "stand[s] a better chance by long distance [telephone] than trying to get him by ham." Transcript in Davidson, "An Instance of Negotiation in a Call Closing" (1978), 124.

47. Hertzberg, *So You Want to be a Ham* (1960), 168.

48. The FCDA exacerbated the position of hams striving to balance federal and neighborhood obligations in the 1950s and 1960s. By taking households as the basic operational unit, civil defense intermingled the priorities of the state and the family. This idea is developed at length in Oakes, *The Imaginary War* (1994); and McEnaney, *Civil Defense Begins at Home* (2000). See also Mechling and Mechling, "The Campaign for Civil Defense" (1991).

6 Ham Radio at Home

1. Carl Dreher and Zeh Bouck, "Our Radio Amateurs," *Harper's Magazine*, October 1941, 535–545, on 545. Portions of this chapter were published previously as Haring, "The 'Freer Men' of Ham Radio" (2003). © Society for the History of Technology. Reprinted with permission of The Johns Hopkins University Press.

2. Dreher and Bouck, "Our Radio Amateurs," 535.

3. Doris S. Couglan, "A Ham's Mother Has Her Say," *QST*, December 1949, 48–49.

4. Joy O. Freed, "He Needs a Hobby," *Parents Magazine*, April 1947, 139–140; Walker A. Tompkins, "An Amazing New Hobby: Ham Radio for the Whole Family," *Parents Magazine*, February 1955, 34–35, 100–102.

5. Marianne Besser, "A Space Program for Your Young Scientist," *Parents Magazine*, January 1962, 48–49, 88; Tompkins, "An Amazing New Hobby"; Wylie, *Generation of Vipers* (1942).

6. Mary Newlin Borton, "Hobbies Can Build Character," *Parents Magazine*, May 1951, 44–46; Besser, "A Space Program for Your Young Scientist."

7. Tompkins, "An Amazing New Hobby"; "The Children's Hour," *Fortune*, August 1954, 68–69; Luciani, *Amateur Radio, Super Hobby* (1980), 39.

8. Carl Dreher and Zeh Bouck, "Our Radio Amateurs," *Harper's Magazine*, October 1941, 535–545, on 545; "Burton New Director, Civil Defense Communications," in "Happenings of the Month," *QST*, December 1950, 25. The announcement of Robert

Burton's appointment included a long personal statement and explained that his reply to a query "for some biographical dope" contained "some youthful experiences of such common amateur interest that we hadn't the heart to cut out even a single word."

9. The origin of this ideology and its propagation through popular culture have been documented in Ehrenreich, *The Hearts of Men* (1983); May, *Homeward Bound* (1988); and Kozol, Life*'s America* (1994).

10. Cuordileone, "Politics in an Age of Anxiety" (2000).

11. Louisa B. Sando, "The YL's Frequency," *CQ*, May 1954, 68, 86, 88–89.

12. Wayne Green, "*CQ*'s Survey," *CQ*, December 1957, 34–35. Allied Radio Corp., *Allied Radio*, catalog no. 150 (1956), 196, 200; and catalog no. 240B (1965), 326. Hertzberg, *So You Want to be a Ham* (1960), 37; Florence V. Collins, "Converting the XYL: New Conversion Data on a Widely-Popular Non-Surplus Item," *CQ*, December 1955, 37–38; Ruth E. Johnson, "Wife's Eye View," *CQ*, July 1946, 18, 60–61. R. W. Johnson, "California Kilowatt," *SCDXC Bulletin*, May 1956; reprinted in *The DXer*, July 1961.

13. On do-it-yourself activity as a hobby, see "Do-It-Yourself: Expected Leisure," in Gelber, *Hobbies* (1999), 268–294. Amelia Black, "The YL's Frequency," *CQ*, June 1946, 40, 42; *The DXer*, July 1961; *The DXer*, June 1957.

14. Station profile of Bob Murphy, *The DXer*, October 1960; Pyle, *Building Up Your Ham Shack* (1960), 83; "Activities," *Rochester DX Association Bulletin*, September 1950; *The DXer*, September 1955; *The DXer*, November/December 1951.

15. May, *Homeward Bound* (1988), 96–97; Epstein, "Anti-Communism, Homophobia, and the Construction of Masculinity" (1994); Costigliola, "Unceasing Pressure for Penetration" (1997); and Cuordileone, "Politics in an Age of Anxiety" (2000).

16. Spigel, *Make Room for TV* (1992), 37–45, 119–127. On the time and class specificity of defining heterosexuality, see Chauncey, *Gay New York* (1994).

17. Baker, *Vox* (1992), 136. Operating ham radio using Morse code afforded anonymity because electrical pulses replaced the human voice. The hobby press recounted cases when male hobbyists pretended to be female in order to play a prank on another hobbyist or simply to attract attention on airwaves overrun with male communicators.

18. Polly Oltion, "The 'We's' Have It," *CQ*, January 1948, 74, 76.

19. Nancy Anderson, "Hobbies," *CQ*, April 1956, 87–88; Ann Gordon, "The Bride," *CQ*, February 1955, 50, 52.

20. Letter quoted in Amelia Black, "The YL's Frequency," *CQ*, April 1947, 52, 67–68; "*CQ* QSL Contest," *CQ*, August 1956, 63. Bauer's card received second prize in a monthly contest for the best confirmation postcard design.

21. Sylvia A. Frank, "Lament of an MYL," *CQ*, February 1949, 24, 77. (Frank proposed "MYL" as a term for a married, non-hobbyist woman, but this usage did not catch on.) National Co. advertisement, *CQ*, December 1953, inside back cover.

22. Florence V. Collins, "Converting the XYL: New Conversion Data on a Widely-Popular Non-Surplus Item," *CQ*, December 1955, 37–38; *The DXer*, May 1956. See also Anne Gardocki, "How to Make Your Wife Love Radio in One Easy Lesson," *CQ*, February 1957, 54–56.

23. James M. Collins, "Nothing... But the Facts," *CQ*, November 1957, 116–117; Maude Phillips quoted in Amelia Black, "The YL's Frequency," *CQ*, June 1946, 40, 42.

24. Pauline Karrol, "Ham Shackles," *CQ*, February 1960, 35, 123–124; "To the Ladies," *The RaRa Rag*, September 1949, 4; Eleanor Wilson, "YL News and Views," *QST*, March 1952, 53, 116, 118; La-RaRa minutes, *The RaRa Rag*, March 1950, 4.

25. "With the Clubs," *The Radio Amateur News*, June 1937, 18–25; *The RaRa Rag*, February 1949, 2; *The DXer*, May 1956 and June 1956; Schenectady Amateur Radio Association, reminder letter about Annual Family Dinner Meeting, 8 May 1959, loose in Box 15, Broughton Papers; "Anaheim Club Will Hold 'No-Speech' Dinner Dance," *Radiogram*, November 1963, 1; *The DXer*, June 1970.

26. Wayne Green, "de [from] W2NSD," *CQ*, October 1955, 11–12, 106, 108, 110, 112–113, on 106; and July 1956, 10.

27. Tom McMullen, "Focus and Comment," *Ham Radio Horizons*, April 1977, 6.

28. From 1950 to 1960, the percentage of married women with children who were part of the paid labor force rose from 19 to 27%. By 1978 this figure reached 50%. Masnick and Bane, *The Nation's Families* (1980), 73.

29. Gail Steckler, "Confessions of an XYL," *CQ*, February 1979, 70.

30. Charlene Knadle, "What's Your ARM-Q?," *CQ*, February 1975, 46–47, 63.

31. Charlene Babb Knadle, "Just Hams," *CQ*, December 1975, 26–28; Steckler, "Confessions of an XYL."

32. "For the XYL's," *The DXer*, September 1972; McCarthy, *CB'ers Guide to Ham Radio* (1979), 136.

33. Ruth E. Johnson, "Wife's Eye View," *CQ*, July 1946, 18, 60–61.

34. Joy O. Freed, "He Needs a Hobby," *Parents Magazine*, April 1947, 139–140; Nancy Anderson, "Hobbies," *CQ*, April 1956, 87–88; quotation from Neil A. Martin, "The Hams Are Still in Business," *Dun's Review*, April 1972, 62–65.

35. Steckler, "Confessions of an XYL."

36. On the centrality of children in the family, see "Baby Boom and Birth Control: The Reproductive Consensus," in May, *Homeward Bound* (1988), 135–161. Constance J. Foster, "The Living Room Belongs to Mother!," *American Home*, April 1943, 28–29.

37. "A Room of his Own: Havens Reflecting Men's Special Interests," *McCall's*, August 1967, 82–87, on 82; Sam Blum, "The Perfect Husband," *McCall's*, August 1967, 60–61, 121–122.

38. Bachelard, *The Poetics of Space* (1957), 18; Wright, "Technology in the Household" (1992), 96.

39. Wright, "Technology in the Household" (1992), 96; Weiss, *History of QRP in the U.S.* (1987), 6; Findlay, *Electronic Experimenter's Manual* (1959), 68. On the popularity since the 1930s of home workshops for do-it-yourself projects, see Gelber, *Hobbies* (1999), 250–251. Gelber also provides, at pp. 204–217, a detailed description of how men claimed a "domestic masculinity" through basement labor. Though his examples are drawn mostly from the first two decades of the twentieth century, Gelber makes clear that the influence of the craftsman era persisted into the 1950s.

40. On appliance aesthetics of the 1950s, see Marling, *As Seen on TV* (1994); and Spigel, *Make Room for TV* (1992), 45–50. Frederick Winsor Jr., "Housing a Hobby: The Residence of Mr. John M. Wells at Southbridge, Massachusetts," *House Beautiful*, February 1933, 36–37, 60–61.

41. Joy O. Freed, "He Needs a Hobby," *Parents Magazine*, April 1947, 139–140; Marianne Besser, "A Space Program for Your Young Scientist," *Parents Magazine*, January 1962, 48–49, 88.

42. Ann Gordon, "The Bride," *CQ*, February 1955, 50, 52; *The DXer*, June 1959.

43. American Radio Relay League, *The Radio Amateur's Handbook* (1955), 481. The quoted phrases remained in the handbook for decades. Weldon Johnson, "QSLs the Photographic Way," *CQ*, September 1957, 42–43, 110.

44. Woolf, *A Room of One's Own* (1929), 23; Lawrence Le Kashman, "Designing the Post-War Ham Shack," *CQ*, September 1946, 30–32. Such sentiments were timeless in the hobby culture. Three decades later a handbook advised, "Every operator's shack is his castle. It ought to be guarded jealously." McCarthy, *CB'ers Guide to Ham Radio* (1979), 304.

45. Woolf, *A Room of One's Own* (1929), 105; "FCC Form 610, Nov. 1955, Application for Amateur Operator and/or Station License," reproduced in Briskman, *Amateur Radio License Guide* (1959), 158; Heathkit/Zenith Educational Systems, *Amateur Radio General License Course* (1979), 10:20.

46. Pyle, *Building Up Your Ham Shack* (1960), 79; Ed Marriner, "A Simple Intercom to the Ham Shack," *CQ*, December 1953, 30.

47. A concise list of the desirable home features repeatedly noted in the hobby literature appears in Karl T. Thurber Jr., "Station Design: How to Plan the Site for Your Ham Station and Set Up Your Equipment," *Ham Radio Horizons*, August 1978, 44–46. Le Kashman, "Designing the Post-War Ham Shack"; Findlay, *Electronic Experimenter's Manual* (1959), 65–79; Pyle, *Building Up Your Ham Shack* (1960), 102–103; American Radio Relay League, *The Radio Amateur's Handbook* (1955), 481.

48. Findlay, *Electronic Experimenter's Manual* (1959), 65; *The DXer*, July 1960; Le Kashman, "Designing the Post-War Ham Shack." Keir Keightley, in "'Turn it down!' she shrieked" (1996), explains that hi-fi hobbyists created sonic separation within shared space by using headphones or by playing music loudly.

49. Nicholas Lefor, "Enclosing the Transmitter," *Radio News*, November 1941, 33, 66; Walter A. Brauer, "The Good Housekeeping Approach to Station Design," *CQ*, August 1949, 15–18, 72. Operating mobile ham radio removed the station from the home but raised aesthetic concerns with regard to the car. Calling it "not too uncommon to find that the XYL [wife] has nearly the last word in the appearance of the family automobile," one hobbyist presented his idea for "A Family Man's Mobile Antenna" that would look normal when mounted on the car. G. Van W. Stivers, "A Family Man's Mobile Antenna," *CQ*, May 1948, 45, 93.

50. Le Kashman, "Designing the Post-War Ham Shack"; Ingram, *Secrets of Ham Radio DXing* (1981), 55, 56; American Radio Relay League, *The Radio Amateur's Handbook* (1955), 481; Pyle, *Building Up Your Ham Shack* (1960), 87, 83.

51. "Harvey-Wells Offers Matching Station," *Radiogram*, April 1956, 1; Berens, *Building the Amateur Radio Station* (1959), 21; Findlay, *Electronic Experimenter's Manual* (1959), 75.

52. From March 1958 to April 1959, *The DXer* did not carry station profiles due to the expense of reproducing the photographs. The shack feature dropped away permanently during a period when the newsletter shrank overall because the club had trouble finding a volunteer editor. *The DXer*, March 1958.

53. Julian N. Jablin, "Picture Your Rig," *CQ*, October 1948, 43, 96; Weldon Johnson, "QSLs the Photographic Way," *CQ*, September 1957, 42–43, 110.

54. Berens, *Building the Amateur Radio Station* (1959), 20; Pyle, *Building Up Your Ham Shack* (1960), 89; Ingram, *Secrets of Ham Radio DXing* (1981), 55; Jack Fulmer, "Clean Up the Shack," *CQ*, October 1956, 74–75, 84; Howard S. Pyle, "Clean Up Your Shack!," *CQ*, April 1965, 43. A non-hobbyist visiting a shack was a captive audience for a lesson on radio procedures. The publisher of a set of forms designed to streamline ham radio paperwork told purchasers that the forms were self explanatory, yet included brief explanations intended to be "interesting to anyone who might visit the 'shack.'" Howard W. Sams and Co., *Amateur Radio Station Manual* (1963), 3.

7 Technical Change and Technical Culture

1. R. Brown, *Electronic Hobbyist's IC Projects Handbook* (1968), 127; Motorola Semiconductor Products, *Integrated Circuit Projects from Motorola* (1966), ii.

2. Morton, *A History of Electronic Entertainment* (1999), 40.

3. Heath Co. president quoted in Lawrence M. Fisher, "Plug Is Pulled on Heathkits, Ending a Do-It-Yourself Era," *New York Times*, 30 March 1992, late edition; Avery Comarow, "A Techie's Toast: 'Here's Looking at You, Kit,'" *Smithsonian*, March 1993, 162; Robert Freed, "On Being an Electronics Buff Today," *Mechanix Illustrated*, March 1975, 60, 111.

4. Bill Orr, "1965–1974: Transistors, FM and Vietnam," *CQ*, January 1995, 100–111, on 109; cartoon reprinted in Beasley, *The Best of Beasley* (1994), 22.

5. Frank Beacham, "Bidding a Fond Farewell to The Do-It-Yourself Heathkits," *Radio World*, 10 June 1992, 15; Perdue, *Heath Nostalgia* (1992), 102; Comarow, "A Techie's Toast."

6. R. Brown, *104 Easy Transistor Projects* (1968), 9; Sam, "Grumbles," *CQ*, March 1968, 63, 116. Joseph O'Connell connected audiophiles' preference for vacuum tubes in the 1980s to the manufacturing of IC-based audio equipment in the Far East and attributed tube nostalgia partly to xenophobia. O'Connell, "The Fine-Tuning of a Golden Ear" (1992). I did not find explicit discussions in the hobby radio literature that would support an analogous conclusion, though nationalism may have factored in hams' technological nostalgia as well.

7. Fisher, "Plug Is Pulled on Heathkits"; Comarow, "A Techie's Toast"; G. Harry Stine, "Life After Heathkits," *Analog Science Fiction and Fact*, July 1993, 214–217.

8. Beacham, "Bidding a Fond Farewell"; Comarow, "A Techie's Toast"; Stine, "Life After Heathkits"; Julian Hirsch, "Heathkit Remembered," *Stereo Review*, June 1996, 38.

9. Allied Radio Corp., *Allied Radio*, catalog no. 240B (1965), front cover; Beason, *How to Build Electronics Kits* (1965), 12–13.

10. On the phenomenon of nostalgia broadly and its particular function in identity construction, see Davis, *Yearning for Yesterday* (1979).

11. "Zero Bias," *CQ*, February 1969, 5; Paul V. Revler and Laurie Margolis, letters to the editor, *CQ*, April 1969, 6.

12. This brief recounting of the history of CB draws on a large number of sources, the most comprehensive of which are Allied Radio Corp., *Understanding and Using Citizens Band Radio* (1963); and Marvin and Schultze, "The First Thirty Years" (1977). A compilation of annual CB license data for 1958–1977 is available in McCarthy, *CB'ers Guide to Ham Radio* (1979), 15. However, only the early figures are useful as operation without a license became more common than with one in the mid 1960s. Estimates for the mid 1970s are from "The Electronic Disease," *Time*, 26 July 1976, 66; and "The Boom in Leisure—Where Americans Spend 160 Billions," *U.S. News and World Report*, 23 May 1977, 62–63. Multiple punctuations and abbreviations of "Citizens' Band" and "Citizens' Banders" were common, sometimes even within the same source, and are retained in quotations here.

13. Comptroller General, *Actions Taken or Needed to Curb Widespread Abuse of the Citizens Band* (1975), i, 6; "Citizens' Band Radio: Trouble in the Air," *Consumer Reports*, October 1974, 744–747.

14. Carl Dreher, "Two-way Radio for Everyman," *Atlantic Monthly*, November 1964, 168, 175–176; Shawn D. Lewis, "10-4, Bro," *Ebony*, October 1976, 120–122, 124, 126. The use of CB radio to organize race-related protests and to strengthen African Americans' social ties also was reported in "The Drivers' Network," *Time*, 22 September 1975, 48–49; and Ronald Ayers, "CB'ers: Riding the Waves," *Essence*, October 1978, 124, 127, 142.

15. McCarthy, *CB'ers Guide to Ham Radio* (1979), vii; Ken Mac Neilage and Dick Malanowicz, letters to the editor, *CQ*, August 1966, 6; "Zero Bias," *CQ*, January 1965, 6.

16. These associations between the ham radio and recreational computing communities particularize general connections observed by other historians. Paul Ceruzzi noted that electronics hobbyists provided "an infrastructure of support" critical to the development of the personal computer. Ceruzzi, "From Scientific Instrument to Everyday Appliance" (1996), 18. Exposure to computers in the workplace has been cited as a source of demand for the personal computer by, among others, Campbell-Kelly and Aspray, *Computer* (1996), 225, 237.

17. Bill Waggoner, letter to the editor, *CQ*, August 1969, 7; Wayne Green, "How *BYTE* Started," *Byte*, September 1975, 9, 96; John Craig, "Editor's Remarks," *Kilobaud*, January 1977, 4–5.

18. Wayne Green, "Publisher's Remarks," *Kilobaud*, January 1977, 3, 14.

19. Member of the Homebrew Computer Club quoted in Levy, *Hackers* (1984), 234.

20. Overbeck and Steffen, *Computer Programs for Amateur Radio* (1984); Anderson, *Computers and the Radio Amateur* (1982), 3, 9; Kasser, *Microcomputers in Amateur Radio* (1981), 9–12, 152–165.

21. "Meeting Notice," *RDXA Bulletin,* April 1982; Overbeck and Steffen, *Computer Programs for Amateur Radio* (1984), 20; Kasser, *Microcomputers in Amateur Radio* (1981), 6–7.

22. Marshall McLuhan attributed "the personal and social consequences of any medium" to "the new scale" introduced by the medium in *Understanding Media* (1964), 7.

23. Pynchon, *The Crying of Lot 49* (1965), 13–14.

References

Full citations to magazine and newspaper articles appear in the Notes and are not duplicated here.

Archives and Special Collections

American Radio Relay League Library, Newington, Connecticut.

William Gundry Broughton Papers. Division of Rare and Manuscript Collections, Cornell University Library, Ithaca, New York.

George H. Clark Radioana Collection, Allen Balcom DuMont Collection, and Paul G. Watson Collection. Archives Center, National Museum of American History, Smithsonian Institution, Washington, DC.

Bruce Kelley Antique Wireless Association Research Center, Bloomfield, New York.

Trade Literature Collection. National Museum of American History Branch Library, Smithsonian Institution Libraries, Washington, DC.

Sources

Adorno, Theodor W. "On the Fetish-Character in Music and the Regression of Listening." 1938. Reprinted in *The Essential Frankfurt School Reader*, ed. Andrew Arato and Eike Gebhardt, 270–299. New York: Continuum, 1988.

Allied Radio Corporation. *Allied Radio*. Merchandise catalogs. Chicago: Allied Radio Corporation. Trade Literature Collection.

———. *Understanding and Using Citizens Band Radio*. Chicago: Allied Radio Corporation, 1963.

American Radio Relay League. *The Radio Amateur's Handbook*. Newington, Connecticut: American Radio Relay League, 1926–1984.

———. *Specialized Communications Techniques for the Radio Amateur*. Newington, Connecticut: American Radio Relay League, 1975.

Anderson, Phil. *Computers and the Radio Amateur*. Englewood Cliffs, New Jersey: Prentice-Hall, Inc., 1982.

The Atomic Cafe. Produced and directed by Jayne Loader, Kevin Rafferty, and Pierce Rafferty. 92 min. The Archives Project, 1982. Videocassette.

Bachelard, Gaston. *The Poetics of Space*. 1957. Translated by Maria Jolas. 1964. Reprint, Boston: Beacon Press, 1994.

Baker, Nicholson. *Vox: A Novel*. New York: Vintage, 1992.

Ballard, David. "Preparing for the Unthinkable War: Civil Defense in the Pacific Northwest during the Cold War Era." *Columbia* 10, no. 3 (fall 1996): 7–13.

Beasley, Robert. *The Best of Beasley*. Sacramento: Worldradio Books, 1994.

Beason, Robert G., ed. *How to Build Electronics Kits: An Easy Guide for Beginners*. Chicago: Allied Radio Corporation, 1965.

Becker, Howard S. *Outsiders: Studies in the Sociology of Deviance*. New York: Free Press, 1963.

Berens, Jack, and Julius Berens. *Building the Amateur Radio Station*. 2nd ed. New York: John F. Rider Publisher, Inc., 1965.

Berens, Julius. *Building the Amateur Radio Station*. New York: John F. Rider Publisher, Inc., 1959.

The Best of Groucho, episode 54-23. John Guedel Productions. Syndicated version of the *You Bet Your Life* episode originally aired on 17 February 1955. Motion Picture and Television Reading Room, Library of Congress.

Bhabha, Homi K. "Interrogating Identity: Frantz Fanon and the postcolonial prerogative." In *The Location of Culture*, 40–65. New York: Routledge, 1994.

Bijker, Wiebe E. *Of Bicycles, Bakelites, and Bulbs: Toward a Theory of Sociotechnical Change*. Cambridge: MIT Press, 1995.

Bijker, Wiebe E., Thomas P. Hughes, and Trevor J. Pinch, eds. *The Social Construction of Technological Systems: New Directions in the Sociology and History of Technology*. Cambridge: MIT Press, 1987.

Bijsterveld, Karin. "'What Do I Do with My Tape Recorder . . . ?': sound hunting and the sounds of everyday Dutch life in the 1950s and 1960s." *Historical Journal of Film, Radio and Television* 24 (2004): 613–634.

Bix, Amy Sue. *Inventing Ourselves Out of Jobs?: America's Debate Over Technological Unemployment, 1929–1981*. Baltimore: Johns Hopkins University Press, 2000.

Blanchard, B. Wayne. "American Civil Defense 1945–1975: The Evolution of Programs and Policies." Ph.D. diss., University of Virginia, 1980.

Boy Scouts of America. *Radio*. Merit Badge Series. New York: Boy Scouts of America, 1930 and 1947.

Boyer, Paul. *By the Bomb's Early Light: American Thought and Culture at the Dawn of the Atomic Age*. New York: Pantheon, 1985.

Braband, Ken C. *The First 50 Years: A History of Collins Radio Company and the Collins Divisions of Rockwell International*. Cedar Rapids, Iowa: Communications Department, Avionics Group, Rockwell International, 1983.

Braden, Donna R. *Leisure and Entertainment in America*. Dearborn, Michigan: Henry Ford Museum and Greenfield Village, 1988.

Briskman, Barry. *Amateur Radio License Guide*. New York: Cowan Publishing, 1959.

Brown, JoAnne. "'A Is for Atom, B Is for Bomb': Civil Defense in American Public Education, 1948–1963." *Journal of American History* 75 (1988): 68–90.

Brown, Robert M. *Electronic Hobbyist's IC Projects Handbook*. Blue Ridge Summit, Pennsylvania: TAB Books, 1968.

———. *104 Easy Projects For the Electronics Gadgeteer*. Blue Ridge Summit, Pennsylvania: TAB Books, 1970.

———. *104 Easy Transistor Projects You Can Build*. Blue Ridge Summit, Pennsylvania: TAB Books, 1968.

———. *104 Simple One-Tube Projects*. Blue Ridge Summit, Pennsylvania: TAB Books, 1969.

Buckwalter, Len. *CB Radio Construction Projects*. Indianapolis: Howard W. Sams and Company, 1963.

———. *Easy-to-Build Transistor Projects*. New Augusta, Indiana: Editors and Engineers, 1965.

———. *The Wonderful, Wacky World of CB Radio*. New York: Grosset and Dunlap, 1976.

Butsch, Richard, ed. *For Fun and Profit: The Transformation of Leisure into Consumption*. Philadelphia: Temple University Press, 1990.

Campbell-Kelly, Martin, and William Aspray. *Computer: A History of the Information Machine*. New York: Basic Books, 1996.

Capstone Electronics Corporation Technical Staff. *How to Use Bargain Transistors.* Cannondale, Connecticut: Capstone Electronics Corporation, 1967.

Caringella, Charles. *Amateur Radio Construction Projects.* Indianapolis: Howard W. Sams and Company, 1963.

———. *Practical Ham Radio Projects.* Indianapolis: Howard W. Sams and Company, 1964.

Carlat, Louis. "'A Cleanser for the Mind': Marketing Radio Receivers for the American Home, 1922–1932." In *His and Hers: Gender, Consumption, and Technology,* ed. Roger Horowitz and Arwen Mohun, 115–137. Charlottesville: University Press of Virginia, 1998.

Carr, Joseph J. *Commodore 64 and 128 Programs for Amateur Radio and Electronics.* Indianapolis: Howard W. Sams, 1986.

———. *The Complete Handbook of Radio Receivers.* Blue Ridge Summit, Pennsylvania: TAB Books, 1980.

Cebik, L. B. *Seven Steps to Designing Your Own Ham Equipment.* Indianapolis: Howard W. Sams and Company, 1979.

Ceruzzi, Paul. "From Scientific Instrument to Everyday Appliance: The Emergence of Personal Computers, 1970–1977." *History and Technology* 13 (1996): 1–31.

Chauncey, George. *Gay New York: Gender, Urban Culture, and the Making of the Gay Male World, 1890–1940.* New York: Basic Books, 1994.

City of Schenectady. *Your City Government, 1941.* Schenectady, New York: City of Schenectady, 1942.

Cockburn, Cynthia. *Brothers: Male Dominance and Technological Change.* London: Pluto, 1983.

Codel, Martin, ed. *Radio and Its Future.* New York: Harper and Brothers, 1930.

Coe, Lewis. *Wireless Radio: A Brief History.* Jefferson, North Carolina: McFarland and Co., 1996.

Cohen, Lizabeth. *Making a New Deal: Industrial Workers in Chicago, 1919–1939.* New York: Cambridge University Press, 1990.

Collins, Francis A. *The Boys' Book of Model Aeroplanes.* New York: The Century Co., 1910.

Collins Radio Company. Promotional materials. Cedar Rapids, Iowa: Collins Radio Company. Trade Literature Collection.

Comptroller General of the United States. United States General Accounting Office. *Actions Taken or Needed to Curb Widespread Abuse of the Citizens Band Radio Service, Federal Communications Commission: Report to the Congress*. Washington, DC: United States General Accounting Office, 1975.

Cones, Harold N., and John H. Bryant. *Zenith Radio The Early Years: 1919–1935*. Atglen, Pennsylvania: Schiffer Publishing, 1997.

Cooper, Constance J., and Ellen Cone. "Defending the Delaware Home Front During World War II." *Delaware History* 26, no. 3–4 (1995–1996): 243–264.

Corn, Joseph J. *The Winged Gospel: America's Romance with Aviation, 1900–1950*. New York: Oxford University Press, 1983.

Corn, Joseph J., and Brian Horrigan. *Yesterday's Tomorrows: Past Visions of the American Future*. Baltimore: Johns Hopkins University Press, 1984.

Costigliola, Frank. "'Unceasing Pressure for Penetration': Gender, Pathology, and Emotion in George Kennan's Formation of the Cold War." *Journal of American History* 83 (1997): 1309–1339.

Cuordileone, K. A. "'Politics in an Age of Anxiety': Cold War Political Culture and the Crisis in American Masculinity, 1949–1960." *Journal of American History* 87 (2000): 515–545.

Cutler, Leon, and Stephen Yura. *A Post-War Career for APO Joe . . . A Million Servicemen Can Create a Billion-Dollar Industry*. New York: Leon Cutler, 1945.

Dachis, Chuck. *Radios By Hallicrafters*. Atglen, Pennsylvania: Schiffer Publishing, 1996.

Dannefer, Dale, and Jill H. Kasen. "Anonymous Exchanges: CB and the Emergence of Sex Typing." *Urban Life* 10 (1981): 265–287.

Dannefer, W. Dale, and Nicholas Poushinsky. "Language and Community." *Journal of Communication* 27 (1977): 122–126.

Dannenmaier, William D. *We Were Innocents: An Infantryman in Korea*. Urbana and Chicago: University of Illinois Press, 1999.

Danzer, Paul, and Richard Roznoy. *Personal Computers in the Ham Shack*. Newington, Connecticut: American Radio Relay League, 1997.

Davidson, Judy. "An Instance of Negotiation in a Call Closing." *Sociology* 12 (1978): 123–133.

Davis, Fred. *Yearning for Yesterday: A Sociology of Nostalgia*. New York: The Free Press, 1979.

de Grazia, Victoria, ed., with Ellen Furlough. *The Sex of Things: Gender and Consumption in Historical Perspective*. Berkeley and Los Angeles: University of California Press, 1996.

de Henseler, Max. *The Hallicrafters Story, 1933–1975*. Charleston, West Virginia: Antique Radio Club of America, 1991.

Dellinger, J. H., and L. E. Whittemore. *Lefax Radio Handbook*. Philadelphia: Lefax, 1922.

DeMaw, Doug. *W1FB's Help for New Hams*. Newington, Connecticut: American Radio Relay League, 1989.

DeSoto, Clinton B. *Two Hundred Meters and Down: The Story of Amateur Radio*. West Hartford, Connecticut: American Radio Relay League, 1936.

Dezettel, Louis M. *Amateur Tests and Measurements*. New Augusta, Indiana: Editors and Engineers, Ltd., 1969.

———. *Realistic Guide to Electronic Kit Building*. Fort Worth: Radio Shack, 1973.

Dr. Strangelove, or How I Learned to Stop Worrying and Love the Bomb. Directed by Stanley Kubrick. 93 min. Hawk Films Ltd., 1964. Videocassette.

Douglas, Alan. *Radio Manufacturers of the 1920s*. 3 vols. Vestal, New York: Vestal Press, 1991.

Douglas, Susan J. "Audio Outlaws: Radio and Phonograph Enthusiasts." In *Possible Dreams: Technological Enthusiasm in America*, ed. John L. Wright, 45–59. Dearborn, Michigan: Henry Ford Museum and Greenfield Village, 1992.

———. *Inventing American Broadcasting, 1899–1922*. Baltimore: Johns Hopkins University Press, 1987.

———. *Listening In: Radio and the American Imagination, from Amos 'n' Andy and Edward R. Murrow to Wolfman Jack and Howard Stern*. New York: Times Books, 1999.

Duston, Merle. *Radio Construction For the Amateur*. 4th ed. Racine, Wisconsin: Whitman Publishing, 1924.

———. *Radio Theory Simplified*. 2nd ed. Racine, Wisconsin: Whitman Publishing, 1926.

Edwards, John. *Exploring Electricity and Electronics with Projects*. Blue Ridge Summit, Pennsylvania: TAB Books, 1983.

Eggers, Dave. *A Heartbreaking Work of Staggering Genius*. New York: Simon and Schuster, 2000.

Ehrenreich, Barbara. *The Hearts of Men: American Dreams and the Flight From Commitment*. Garden City, New York: Anchor Press/Doubleday, 1983.

The EKKO Broadcasting Station Stamp Album: Containing Spaces for Stamps for Every Broadcasting Station in the United States and Canada. Chicago: EKKO, 1924.

Epstein, Barbara. "Anti-Communism, Homophobia, and the Construction of Masculinity in the Postwar U.S." *Critical Sociology* 20 (1994): 21–44.

Evenson, R. C., and O. R. Beach. *Surplus Radio Conversion Manual*. 2 vols. Summerland, California: Editors and Engineers, Ltd., 1948.

Farman, Irvin. *Tandy's Money Machine: How Charles Tandy Built Radio Shack Into the World's Largest Electronics Chain*. Chicago: Mobius Press, 1992.

Ferguson, Eugene S. *Engineering and the Mind's Eye*. Cambridge: MIT Press, 1992.

Ferrell, Nancy Warren. *The New World of Amateur Radio*. New York: Franklin Watts, 1986.

Findlay, David A. *The Electronic Experimenter's Manual*. New York: Ziff-Davis, 1959.

Fischer, Claude S. "Changes in Leisure Activities, 1890–1940." *Journal of Social History* 27 (1994): 453–475.

Fisk, James R., ed. *Ham Notebook*. Greenville, New Hampshire: Communications Technology, 1973.

Frequency. Directed by Gregory Hoblit. 118 min. New Line Cinema, 2000. Videocassette.

Fried, Richard M. *The Russians Are Coming! The Russians Are Coming!: Pageantry and Patriotism in Cold-War America*. New York: Oxford University Press, 1998.

Frye, John T. *Radio Receiver Servicing*. Indianapolis: Howard W. Sams and Company, 1955.

Garrison, Dee. "'Our Skirts Gave them Courage': The Civil Defense Protest Movement in New York City, 1955–1961." In *Not June Cleaver: Women and Gender in Postwar America, 1945–1960*, ed. Joanne Meyerowitz, 201–226. Philadelphia: Temple University Press, 1994.

Gelber, Steven M. "Do-It-Yourself: Constructing, Repairing and Maintaining Domestic Masculinity." *American Quarterly* 49 (1997): 66–112.

———. *Hobbies: Leisure and the Culture of Work in America*. New York: Columbia University Press, 1999.

———. "A Job You Can't Lose: Work and Hobbies in the Great Depression." *Journal of Social History* 24 (1991): 741–766.

General Electric Company. *Annual Report.* Schenectady, New York: General Electric, 1945–1960.

———. Radio Department. *Radio Operators' Manual.* 3rd ed. Schenectady, New York: General Electric Company, 1938.

Gitelman, Lisa. *Scripts, Grooves, and Writing Machines: Representing Technology in the Edison Era.* Stanford: Stanford University Press, 1999.

———. "Souvenir Foils: On the Status of Print at the Origin of Recorded Sound." In *New Media, 1740–1915,* ed. Lisa Gitelman and Geoffrey B. Pingree, 157–173. Cambridge: MIT Press, 2003.

Gitelman, Lisa, and Geoffrey B. Pingree, eds. *New Media, 1740–1915.* Cambridge: MIT Press, 2003.

Gleason, William A. *The Leisure Ethic: Work and Play in American Literature.* Stanford: Stanford University Press, 1999.

Godley, Paul. *Getting Acquainted with Radio Receivers.* Upper Montclair, New Jersey: Adams-Morgan Company, 1923.

Grammer, George. *Understanding Amateur Radio.* 2nd ed. Newington, Connecticut: American Radio Relay League, 1976.

Gregory, Danny, and Paul Sahre. *Hello World: A Life in Ham Radio.* New York: Princeton Architectural Press, 2003.

Grover, Kathryn, ed. *Hard at Play: Leisure in America, 1840–1940.* Amherst: University of Massachusetts Press, 1992.

Grubbs, Jim. *The Commodore Ham's Companion.* Springfield, Illinois: QSKY Publishing, 1985.

Hall, A. Neely. *Home Handicraft For Boys: Learning Through Doing.* 2nd ed. Philadelphia: J. B. Lippincott, 1935.

Hallicrafters Company. *Annual Report.* Chicago: Hallicrafters Company, 1950.

———. *Hallicrafters Radio.* Merchandise catalogs. Chicago: Hallicrafters Company. Trade Literature Collection.

Halprin, Robert J. *The Code Book: Amateur Radio CW Operating.* With a foreword by George Hart. Lake Geneva, Wisconsin: Tiare Publications, 1993.

Hamburg, Merrill. *Beginning to Fly: The Book of Model Airplanes.* Boston: Houghton Mifflin, 1928.

Hammarlund Manufacturing Company. Promotional materials. New York: Hammarlund Manufacturing Company. Trade Literature Collection.

Haring, Kristen. "The 'Freer Men' of Ham Radio: How a Technical Hobby Provided Social and Spatial Distance." *Technology and Culture* 44 (2003): 734–761.

———. "Technical Identity in the Age of Electronics." Ph.D. diss., Harvard University, 2002.

Harper, Douglas A. *Working Knowledge: Skill and Community in a Small Shop*. Chicago: University of Chicago Press, 1987.

Hart, George. *A Manual for the War Emergency Radio Service*. West Hartford, Connecticut: American Radio Relay League, 1944.

Headrick, Daniel R. *The Invisible Weapon: Telecommunications and International Politics, 1851–1945*. New York: Oxford University Press, 1991.

Heath Company. Heathkit assembly manuals. Benton Harbor, Michigan: Heath Company. Private collection.

Heathkit/Zenith Educational Systems. *Amateur Radio General License Course*. 2 vols. Benton Harbor, Michigan: Heath Company, 1979.

Helfrick, Albert D. *Amateur Radio Equipment Fundamentals*. Englewood Cliffs, New Jersey: Prentice-Hall, 1982.

Henriksen, Margot A. *Dr. Strangelove's America: Society and Culture in the Atomic Age*. Berkeley and Los Angeles: University of California Press, 1997.

Hertzberg, Robert. *So You Want to be a Ham*. 2nd ed. Indianapolis: Howard W. Sams and Company, 1960.

Hevly, Bruce. "The Tools of Science: Radio, Rockets, and the Science of Naval Warfare." In *National Military Establishments and the Advancement of Science and Technology: Studies in 20th Century History*, ed. Paul Forman and José M. Sánchez-Ron, 215–232. Boston: Kluwer Academic Publishers, 1996.

Hilmes, Michele. *Radio Voices: American Broadcasting, 1922–1952*. Minneapolis: University of Minnesota Press, 1997.

Horowitz, Roger, ed. *Boys and Their Toys?: Masculinity, Technology, and Class in America*. New York: Routledge, 2001.

Horowitz, Roger, and Arwen Mohun, eds. *His and Hers: Gender, Consumption, and Technology*. Charlottesville: University Press of Virginia, 1998.

Howard W. Sams and Company. *Amateur Radio Station Manual*. Indianapolis: Howard W. Sams and Company, 1963.

Howeth, L. S. *History of Communications-Electronics in the United States Navy*. Washington, DC: United States Government Printing Office, 1963.

Hydro-Aire. Electronics Division. *The Transistor and You: A Handbook for Radio and Electronics Amateurs on Semiconductor Devices and their Applications*. Burbank, California: Hydro-Aire, 1955.

Ingram, Dave. *44 Electronics Projects for Hams, SWLs, CBers & Radio Experimenters*. Blue Ridge Summit, Pennsylvania: TAB Books, 1981.

———. *Secrets of Ham Radio DXing*. Blue Ridge Summit, Pennsylvania: TAB Books, 1981.

Jaker, Bill, Frank Sulek, and Peter Kanze. *The Airwaves of New York: Illustrated Histories of 156 AM Stations in the Metropolitan Area, 1921–1996*. Jefferson, North Carolina: McFarland and Company, 1998.

Jenkins, Reese V. "Technology and the Market: George Eastman and the Origins of Mass Amateur Photography." *Technology and Culture* 16 (1975): 1–19.

Jome, Hiram L. *Economics of the Radio Industry*. Chicago: A. W. Shaw Company, 1925.

Jones, Bernard E., ed. *The Encyclopaedia of Early Photography*. 1911. Reprint, London: Bishopsgate Press, 1981.

Kasser, Joe. *Microcomputers in Amateur Radio*. Blue Ridge Summit, Pennsylvania: TAB Books, 1981.

Kearman, Jim. *Low Profile Amateur Radio: Operating a Ham Station From Almost Anywhere*. Newington, Connecticut: American Radio Relay League, 1993.

Keightley, Keir. "'Turn it down!' she shrieked: gender, domestic space, and high fidelity, 1948–1959." *Popular Music* 15 (1996): 149–177.

Kittler, Friedrich A. *Gramophone, Film, Typewriter*. 1986. Translated by Geoffrey Winthrop-Young and Michael Wutz. Stanford: Stanford University Press, 1999.

Kline, Ronald, and Trevor Pinch. "Users as Agents of Technological Change: The Social Construction of the Automobile in the Rural United States." *Technology and Culture* 37 (1996): 763–795.

Kneitel, Thomas S., ed. *Jobs and Careers in Electronics 1961*. "A *Popular Electronics* Annual." New York: Ziff-Davis, 1960.

Kornfeld, Lewis F. *President's Messages: A Collection of Messages by Lewis F. Kornfeld Addressed to Employees and Customers of Radio Shack*. Fort Worth, Texas: Tandy Corporation, 1981.

———. *To Catch a Mouse, Make a Noise Like a Cheese*. Englewood Cliffs, New Jersey: Prentice-Hall, 1983.

Kozol, Wendy. Life*'s America: Family and Nation in Postwar Photojournalism*. Philadelphia: Temple University Press, 1994.

Kuslan, Richard David, and Louis I. Kuslan. *Ham Radio: An Introduction to the World Beyond CB*. Englewood Cliffs, New Jersey: Prentice Hall, 1981.

Lamson, Robert. "The Army and Civil Defense." *Military Review* 44 (1964): 3–12.

LaVoy, Kenneth R. *Problems and Projects in Industrial Arts*. Peoria, Illinois: The Manual Arts Press, 1924.

Levey, Alex. *Radio Receiver Laboratory Manual*. New York: John F. Rider Publisher, 1956.

Levy, Steven. *Hackers: Heroes of the Computer Revolution*. Garden City, New York: Anchor Press/Doubleday, 1984.

Light, Jennifer S. "When Computers Were Women." *Technology and Culture* 40 (1999): 455–483.

Lindemann, Carl, Jr. "Marketing Communications Receivers: A Study of the Marketing Potential for Communications Type Radio Receivers in New Jersey, Massachusetts, and Kentucky for the Year 1947." Manuscript prepared for a class in the Department of Business and Engineering Administration, Massachusetts Institute of Technology, 1947. American Radio Relay League Library.

Lindsay, Christina. "From the Shadows: Users as Designers, Producers, Marketers, Distributors, and Technical Support." In *How Users Matter: The Co-Construction of Users and Technologies*, ed. Nelly Oudshoorn and Trevor Pinch, 29–50. Cambridge: MIT Press, 2003.

Luciani, Vince. *Amateur Radio, Super Hobby!: What it Is, Who We Are, How to Join*. Cologne, New Jersey: Cologne Press, 1980.

Lytel, Allan H. *Two-Way Radio*. New York: McGraw-Hill, 1959.

Macay, E. G. *Radio How and Why*. Philadelphia: Radio Hand Book Publishing Company, 1922.

MacDonald, J. Fred. *Television and the Red Menace: The Video Road to Vietnam*. New York: Praeger, 1985.

Mannes, George. "The Birth of Cable TV." *Invention and Technology* 12, no. 2 (fall 1996): 42–50.

Marling, Karal Ann. *As Seen on TV: The Visual Culture of Everyday Life in the 1950s*. Cambridge: Harvard University Press, 1994.

Marsden, Peter V., John Shelton Reed, Michael D. Kennedy, and Kandi M. Stinson. "American Regional Cultures and Differences in Leisure Time Activities." *Social Forces* 60 (1982): 1023–1049.

Marsh, Margaret. "Suburban Men and Masculine Domesticity, 1870–1915." *American Quarterly* 40 (1988): 165–186.

Marten, James. "Coping with the Cold War: Civil Defense in Austin, Texas, 1961–1962." *East Texas Historical Journal* 26 (1988): 3–13.

Martin, Michèle. *"Hello, Central?": Gender, Technology, and Culture in the Formation of Telephone Systems*. Montreal: McGill/Queen's University Press, 1991.

Marvin, Carolyn. *When Old Technologies Were New: Thinking About Electric Communication in the Late Nineteenth Century*. New York: Oxford University Press, 1988.

Marvin, Carolyn, and Quentin J. Schultze. "The First Thirty Years." *Journal of Communication* 27 (1977): 104–117.

Masnick, George, and Mary Jo Bane. *The Nation's Families: 1960–1990*. Cambridge: Joint Center for Urban Studies of MIT and Harvard University, 1980.

Matthews, Blayney F. *The Specter of Sabotage*. Los Angeles: Lymanhouse, 1941.

May, Elaine Tyler. *Homeward Bound: American Families in the Cold War Era*. New York: Basic Books, 1988.

McCarthy, George W. *CB'ers Guide to Ham Radio*. New York: Van Nostrand Reinhold, 1979.

McCarthy, John, and Peter Wright. *Technology as Experience*. Cambridge: MIT Press, 2004.

McEnaney, Laura. *Civil Defense Begins at Home: Militarization Meets Everyday Life in the Fifties*. Princeton: Princeton University Press, 2000.

McGirr, Lisa. *Suburban Warriors: The Origins of the New American Right*. Princeton: Princeton University Press, 2001.

McIlroy, Andrew. "No Interest, No Time, No Money: Civil Defense in Cleveland in the Cold War." *Ohio History* 106 (1997): 59–86.

McLuhan, Marshall. *Understanding Media: The Extensions of Man*. 1964. Reprint, Cambridge: MIT Press, 1994.

McPherson, Marti, and Forest H. Belt. *Forest H. Belt's Easi-Guide to CB Radio for the Family*. Indianapolis: Howard W. Sams and Company, 1975.

Mechling, Elizabeth Walker, and Jay Mechling. "The Campaign for Civil Defense and the Struggle to Naturalize the Bomb." *Western Journal of Speech Communication* 55 (1991): 105–133.

Mensel, Robert E. "'Kodakers Lying in Wait': Amateur Photography and the Right of Privacy in New York, 1885–1915." *American Quarterly* 43 (1991): 24–45.

Meyerowitz, Joanne. "Beyond the Feminine Mystique: A Reassessment of Postwar Mass Culture, 1946–1958." *Journal of American History* 79 (1993): 1455–1482.

Millen, James, ed. *Radio Design Practice.* Malden, Massachusetts: James Millen, Inc., 1935.

Millen, James, and Robert S. Kruse, eds. *Below Ten Meters: The Manual of Ultra-Short-Wave-Radio.* Malden, Massachusetts: The National Company, 1932.

Miller, Jay H. *The Pocket Guide to Collins Amateur Radio Equipment, 1946 to 1980.* Dallas: Trinity Graphics Systems, 1995.

Miller, Robert Earnest. "The War that Never Came: Civilian Defense in Cincinnati, Ohio, During World War II." *Queen City Heritage: The Journal of the Cincinnati Historical Society* 49, no. 4 (winter 1991): 2–22.

Mills, C. Wright. *White Collar: The American Middle Classes.* New York: Oxford University Press, 1951.

Mims, Forrest M., III. *Integrated Circuit Projects.* 6 vols. Fort Worth, Texas: Radio Shack, 1973–1977.

Minnesota Office of Civilian Defense. Welfare Division. *Victory Aide Handbook.* Saint Paul: Minnesota Office of Civilian Defense, [1942 or 1943].

Morton, David L. *A History of Electronic Entertainment Since 1945.* New Brunswick, New Jersey: IEEE History Center, 1999.

Motorola Semiconductor Products. *Integrated Circuit Projects from Motorola.* Phoenix: Motorola, 1966.

New Jersey Defense Council. *The Control System of the Civil Protection Services.* Trenton, New Jersey: New Jersey Defense Council, 1942.

New York State Civil Defense Commission. *Radio Officer's Guide: Radio Amateur Civil Emergency Service.* New York: New York State Civil Defense Commission, 1961.

Nickles, Shelley Kaplan. "Object Lessons: Household Appliance Design and the American Middle Class, 1920–1960." Ph.D. diss., University of Virginia, 1999.

Noble, David F. *Forces of Production: A Social History of Industrial Automation.* New York: Knopf, 1984.

Nye, David E. *Electrifying America: Social Meanings of a New Technology.* Cambridge: MIT Press, 1990.

Nye, Miriam Baker. "George R. Call: A Pioneer in Ham Radio." *Annals of Iowa* 42 (1974): 478–483.

Oakes, Guy. *The Imaginary War: Civil Defense and American Cold War Culture.* New York: Oxford University Press, 1994.

O'Connell, Joseph. "The Fine-Tuning of a Golden Ear: High-End Audio and the Evolutionary Model of Technology." *Technology and Culture* 33 (1992): 1–37.

Oldenziel, Ruth. "Boys and Their Toys: The Fisher Body Craftsman's Guild, 1930–1968, and the Making of a Male Technical Domain." *Technology and Culture* 38 (1997): 60–96.

———. *Making Technology Masculine: Men, Women and Modern Machines in America, 1870–1945.* Amsterdam: Amsterdam University Press, 1999.

Orr, William I. *Better Shortwave Reception: A Handbook for the Radio Amateur and Shortwave Listener.* Wilton, Connecticut: Radio Publications, Inc., 1957.

Otis, Laura. "The Other End of the Wire: Uncertainties of Organic and Telegraphic Communication." *Configurations* 9 (2001): 181–206.

Oudshoorn, Nelly, and Trevor Pinch, eds. *How Users Matter: The Co-Construction of Users and Technologies.* Cambridge: MIT Press, 2003.

Overbeck, Wayne, and James A. Steffen. *Computer Programs for Amateur Radio.* Hasbrouck Heights, New Jersey: Hayden, 1984.

Parkinson, Ethelyn M. *Today I Am a Ham.* New York: Abingdon Press, 1968.

Peiss, Kathy. *Cheap Amusements: Working Women and Leisure in Turn-of-the-Century New York.* Philadelphia: Temple University Press, 1986.

———. *Hope in a Jar: The Making of America's Beauty Culture.* New York: Metropolitan Books, 1998.

Perdue, Terry. *Heath Nostalgia.* Lynnwood, Washington: Terry Perdue, 1992.

Perlman, Marc. "Golden Ears and Meter Readers: The Contest for Epistemic Authority in Audiophilia." *Social Studies of Science* 34 (2004): 783–807.

Perrucci, Robert, and Joel E. Gerstl, eds. *The Engineers and the Social System.* New York: John Wiley and Sons, 1969.

Peterson, Richard A. "Measuring Culture, Leisure, and Time Use." *Annals of the American Academy of Political and Social Science* 453 (1981): 169–179.

Pinch, Trevor, and Frank Trocco. *Analog Days: The Invention and Impact of the Moog Synthesizer.* Cambridge: Harvard University Press, 2002.

Pollack, Harvey. *Transistor Theory and Circuits Made Simple.* New York: American Electronics Company, 1958.

Post, Robert C. *High Performance: The Culture and Technology of Drag Racing, 1950–1990*. Baltimore: Johns Hopkins University Press, 1994.

Powell, A. M. *Home-made Electrical Apparatus: A Practical Handbook for Amateur Experimenters*. 3 vols. New York: Cole and Morgan, 1917–1918.

Prentiss, Augustin M. *Civil Defense in Modern War*. New York: McGraw-Hill, 1951.

Proceedings of General Electric Radio Specialists' Meeting. Confidential publication, copy number 34 of 125. Schenectady, New York: General Electric Company Radio and Television Department, 1940. Broughton Papers.

Pyle, Howard S. *ABC's of Ham Radio*. 2nd ed. Indianapolis: Howard W. Sams and Company, 1965.

———. *Building Up Your Ham Shack*. Indianapolis: Howard W. Sams and Company, 1960.

Pynchon, Thomas. *The Crying of Lot 49*. 1965. Reprint, New York: Perennial Classics, 1999.

Radio Club of America. *Yearbook for 1930*. New York: Radio Club of America, 1930.

Radio Corporation of America. *Annual Report*. New York: RCA, 1927–1960.

———. Department of Information. *What's the Right Word?: A Dictionary of Common and Uncommon Terms in Radio, Television, Electronics*. New York: Radio Corporation of America, 1952.

———. RCA Victor Division. *Common Words in Radio, Television and Electronics*. Camden, New Jersey: Radio Corporation of America, 1947.

Riley, Michael R., ed. *The Emergency Coordinator's Handbook*. Newington, Connecticut: American Radio Relay League, 1984.

Rodger, Andrew C. "So Few Earnest Workers, 1914–1930." In *Private Realms of Light: Amateur photography in Canada, 1839–1940*, ed. Lilly Koltun, 72–87. Markham, Ontario: Fitzhenry and Whiteside, 1984.

Rose. *Speak to the World: Amateur Radio Language Guide*. Mundelein, Illinois: Rose, 1991.

Rosenzweig, Roy. *Eight Hours for What We Will: Workers and Leisure in an Industrial City, 1870–1920*. Cambridge: Cambridge University Press, 1983.

Schatzberg, Eric. *Wings of Wood, Wings of Metal: Culture and Technical Choice in American Airplane Materials, 1914–1945*. Princeton: Princeton University Press, 1999.

Scheer, Robert. *With Enough Shovels: Reagan, Bush, and Nuclear War*. New York: Random House, 1982.

Schiffer, Michael Brian. "Cultural Imperatives and Product Development: The Case of the Shirt-Pocket Radio." *Technology and Culture* 34 (1993): 98–113.

Schrecker, Ellen. *The Age of McCarthyism: A Brief History with Documents.* The Bedford Series in History and Culture. New York: Bedford Books, 1994.

Scipione, Paul A. *M.A.R.S.: Calling Back To 'The World' From Vietnam.* Kalamazoo, Michigan: The Center for the Study of the Vietnam War, 1994.

Shuart, George W. *Radio Amateur Course: Including Constructional Articles on Transmitters and Receivers for HAM Use.* New York: Short Wave and Television Magazine, 1937.

Sienkiewicz, Julian M. *Vacuum Tube Circuits for the Electronic Experimenter.* New York: Ziff-Davis, 1961.

Skolnik, Richard. *The Wonderful World of Ham Radio: An Introduction for Young People.* N.p.: MFJ Enterprises, Inc., 1990.

Smith, J. Jerome. "Gender Marking on Citizens Band Radio: Self-Identity in a Limited-Channel Speech Community." *Sex Roles* 7 (1981): 599–606.

Smith, Michael L. "Selling the Moon: The U.S. Manned Space Program and the Triumph of Commodity Scientism." In *The Culture of Consumption: Critical Essays in American History, 1880–1980,* ed. Richard Wightman Fox and T. J. Jackson Lears, 175–209. New York: Pantheon Books, 1983.

Smith-Divita, Stephanie. "Electrical Men: The Electrical League of Cleveland, 1909–1949, and Selling Power." Ph.D. diss., Case Western Reserve University, 2000.

Snyder, Thomas S., ed. *Air Force Communications Command: 1938–1991, An Illustrated History.* Scott Air Force Base, Illinois: AFCC Office of History, 1991.

Sobell, Morton. *On Doing Time.* New York: Charles Scribner's Sons, 1974.

Spain, Daphne. *Gendered Spaces.* Chapel Hill: University of North Carolina Press, 1992.

Spigel, Lynn. *Make Room for TV: Television and the Family Ideal in Postwar America.* Chicago: University of Chicago Press, 1992.

The Standard Periodical Directory, 1964–1965. New York: Oxbridge Publishing, 1964.

Stanford Research Institute. *Amateur Radio: An International Resource for Technological, Economic, and Sociological Development.* Menlo Park, California: SRI, 1966.

Stebbins, Robert A. *Amateurs: On the Margin Between Work and Leisure.* Beverly Hills: Sage Publications, 1979.

———. *Amateurs, Professionals, and Serious Leisure.* Montreal and Kingston: McGill-Queen's University Press, 1992.

———. *New Directions in the Theory and Research of Serious Leisure.* Mellen Studies in Sociology, vol. 28. Lewiston, New York: Edwin Mellen Press, 2001.

Sterling, George E. *The Radio Manual: For Radio Engineers, Inspectors, Students, Operators and Radio Fans.* New York: D. Van Nostrand Company, 1928.

Stoner, Donald L. *Transistor Transmitters for the Amateur.* Indianapolis: Howard W. Sams and Company, 1964.

Stoner, Donald L., and L. A. Earnshaw. *The Transistor Radio Handbook: Theory, Circuitry, Equipment.* Summerland, California: Editors and Engineers, Ltd., 1963.

Stong, C. L. *The* Scientific American *Book of Projects for the Amateur Scientist.* With an introduction by Vannevar Bush. New York: Simon and Schuster, 1960.

Strom, Dave. *Power Up!: How to Make Battery Adapters for Portable and Mobile Military Radios, and Other Military and Non-Military Electronics.* Commack, New York: CRB Research Books, Inc., 1994.

Stuit, Dewey B., ed. *Personnel Research and Test Development in the Bureau of Naval Personnel.* Princeton: Princeton University Press, 1947.

Swan, Franklin, and Warren E. Palmer, Jr. *Practical Electronics Series.* Vol. 2, *Construction Techniques and Projects.* Fort Worth: Radio Shack, 1977.

Swiencicki, Mark A. "Consuming Brotherhood: Men's Culture, Style and Recreation as Consumer Culture, 1880–1930." *Journal of Social History* 31 (1998): 773–808.

TAB Editorial Staff, eds. *CB Radio Schematic/Servicing Manual.* 4 vols. Blue Ridge Summit, Pennsylvania: TAB Books, 1976.

Taylor, Dorceta E. *Identity in Ethnic Leisure Pursuits.* San Francisco: Mellen Research University Press, 1992.

Technical Advertising Associates. *Radio Component Handbook.* Cheltenham, Pennsylvania: Technical Advertising Associates, 1948.

Tester, Jerry Virgil. "The Effect of Amateur Radio on Student Achievement in a Beginning College Level Electronics Course." Ph.D. diss., Texas A&M University, 1977.

Tompkins, Walker A. *CQ Ghost Ship!* Philadelphia: Macrae Smith Company, 1960.

———. *DX Brings Danger.* Santa Barbara: Sagamore Books, 1962.

———. *SOS at Midnight.* Santa Barbara: Sagamore Books, 1957.

Transistor Specifications and Substitution Handbook. Brownsburg, Indiana: Tech Press, Inc., 1966.

Turkle, Sherry. *Life on the Screen: Identity in the Age of the Internet.* New York: Touchstone, 1995.

———. *The Second Self: Computers and the Human Spirit.* New York: Simon and Schuster, 1984.

Turner, Rufus P. *Integrated Circuits: Fundamentals and Projects.* Chicago: Allied Radio Corporation, 1968.

United States Department of Commerce. *Commerce Yearbook 1932.* Washington, DC: United States Government Printing Office, 1932.

———. Bureau of the Census. *Historical Statistics of the United States, Colonial Times to 1970.* Washington, DC: United States Government Printing Office, 1975.

———. Radio Division. *Amateur Radio Stations of the United States.* Washington, DC: United States Government Printing Office, 1931.

United States Department of Labor. Bureau of Labor Statistics. *Employment Outlook and Changing Occupational Structure in Electronics Manufacturing.* Bulletin No. 1363. Washington, DC: United States Government Printing Office, 1963.

———. *Employment Outlook for Engineers.* Bulletin No. 968. Washington, DC: United States Government Printing Office, 1950.

———. *Employment Outlook in Electronics Manufacturing.* Bulletin No. 1072. Washington, DC: United States Government Printing Office, 1952.

———. *Occupational Outlook Handbook: Employment Information on Major Occupations for Use in Guidance.* Bulletin No. 940. Washington, DC: United States Government Printing Office, 1948.

United States Federal Civil Defense Administration. *Annotated Civil Defense Bibliography for Teachers.* Rev. ed. Washington, DC: United States Government Printing Office, 1956.

———. *Annual Report.* Washington, DC: United States Government Printing Office, 1951–1958.

———. *Battleground U.S.A.: An Operations Plan for the Civil Defense of a Metropolitan Target Area.* Washington, DC: United States Government Printing Office, 1957.

———. *National Civil Defense Conference Report.* Washington, DC: United States Government Printing Office, [1951].

United States Federal Communications Commission. *Annual Report.* Washington, DC: United States Government Printing Office, 1935–1994.

United States Federal Radio Commission. *Annual Report.* Washington, DC: United States Government Printing Office, 1927–1934.

United States National Security Resources Board. *United States Civil Defense.* Washington, DC: United States Government Printing Office, 1950.

United States Office of Civil and Defense Mobilization. *Annual Report*. Washington, DC: United States Government Printing Office, 1959–1961.

———. *Annual Statistical Report*. Washington, DC: United States Government Printing Office, 1959–1961.

United States Office of Civil Defense. *Annual Report*. Washington, DC: United States Government Printing Office, 1962–1971.

———. *1964 Annual Statistical Report*. Washington, DC: United States Government Printing Office, 1965.

United States Office of Civil Defense Planning. *Civil Defense for National Security* (the Hopley Report). Washington, DC: United States Government Printing Office, 1948.

United States Office of Civilian Defense. *A Handbook for Messengers*. Washington, DC: United States Government Printing Office, 1942.

United States War Production Board. *Durable Goods—Second and Third Consumer Requirements Surveys, March and April, 1944*. Washington, DC: War Production Board, 1944.

Vanek, Joann. "Work, Leisure, and Family Roles: Farm Households in the United States, 1920–1955." *Journal of Family History* 5 (1980): 422–431.

Wall, Cynthia. *Disappearing Act*. Salem, Oregon: Dimi Press, 1996.

———. *Firewatch!* Newington, Connecticut: American Radio Relay League, 1993.

Weart, Spencer R. *Nuclear Fear: A History of Images*. Cambridge: Harvard University Press, 1988.

Weiss, Adrian. *History of QRP in the U.S., 1924–1960*. Vermillion, South Dakota: Milliwatt Books, 1987.

Wels, Byron. *Here is Your Hobby: Amateur Radio*. New York: G. P. Putnam's Sons, 1968.

Whyte, William H., Jr. *The Organization Man*. New York: Simon and Schuster, 1956.

Wiebe, Robert H. *The Search for Order, 1877–1920*. New York: Hill and Wang, 1967.

Wilber, Gordon O., and Emerson E. Neuthardt. *Aeronautics in the Industrial Arts Program: A Handbook for Students and Teachers*. Air-Age Education Series. New York: Macmillan Company, 1942.

Winkler, Allan M. "A 40-year history of civil defense." *Bulletin of the Atomic Scientists* 40, no. 6 (June/July 1984): 16–22.

Winter, William. *The Model Aircraft Handbook*. New York: Thomas Y. Crowell Co., 1942.

Wood, Harry E., and James H. Smith. *Prevocational and Industrial Arts*. Chicago: Atkinson, Mentzer and Co., 1919.

Woolf, Virginia. *A Room of One's Own*. 1929. Reprint, with a foreword by Mary Gordon, New York: Harcourt Brace and Company, 1981.

Wright, John L. "Technology in the Household: A Reminiscence." In *Possible Dreams: Technological Enthusiasm in America*, ed. John L. Wright, 94–97. Dearborn, Michigan: Henry Ford Museum and Greenfield Village, 1992.

Wylie, Philip. *Generation of Vipers*. New York: Rinehart, 1942.

Yates, Raymond Francis, ed. *Everyman's Guide to Radio: A Practical Course of Common-Sense Instruction in The World's Most Fascinating Science*. 4 vols. New York: Popular Radio, Inc., 1926.

Zimmermann, Patricia R. *Reel Families: A Social History of Amateur Film*. Bloomington and Indianapolis: Indiana University Press, 1995.

Index

Series List

Inside Technology

edited by Wiebe E. Bijker, W. Bernard Carlson, and Trevor Pinch

Janet Abbate, *Inventing the Internet*

Charles Bazerman, *The Languages of Edison's Light*

Marc Berg, *Rationalizing Medical Work: Decision-Support Techniques and Medical Practices*

Wiebe E. Bijker, *Of Bicycles, Bakelites, and Bulbs: Toward a Theory of Sociotechnical Change*

Wiebe E. Bijker and John Law, editors, *Shaping Technology/Building Society: Studies in Sociotechnical Change*

Stuart S. Blume, *Insight and Industry: On the Dynamics of Technological Change in Medicine*

Pablo J. Boczkowski, *Digitizing the News: Innovation in Online Newspapers*

Geoffrey C. Bowker, *Memory Practices in the Sciences*

Geoffrey C. Bowker, *Science on the Run: Information Management and Industrial Geophysics at Schlumberger, 1920–1940*

Geoffrey C. Bowker and Susan Leigh Star, *Sorting Things Out: Classification and Its Consequences*

Louis L. Bucciarelli, *Designing Engineers*

H. M. Collins, *Artificial Experts: Social Knowledge and Intelligent Machines*

Paul N. Edwards, *The Closed World: Computers and the Politics of Discourse in Cold War America*

Herbert Gottweis, *Governing Molecules: The Discursive Politics of Genetic Engineering in Europe and the United States*

Kristen Haring, *Ham Radio's Technical Culture*

Gabrielle Hecht, *The Radiance of France: Nuclear Power and National Identity after World War II*

Kathryn Henderson, *On Line and On Paper: Visual Representations, Visual Culture, and Computer Graphics in Design Engineering*

Anique Hommels, *Unbuilding Cities: Obduracy in Urban Sociotechnical Change*

David Kaiser, editor, *Pedagogy and the Practice of Science: Historical and Contemporary Perspectives*

Peter Keating and Alberto Cambrosio, *Biomedical Platforms: Reproducing the Normal and the Pathological in Late-Twentieth-Century Medicine*

Eda Kranakis, *Constructing a Bridge: An Exploration of Engineering Culture, Design, and Research in Nineteenth-Century France and America*

Christophe Lécuyer, *Making Silicon Valley: Innovation and the Growth of High Tech, 1930–1970*

Pamela E. Mack, *Viewing the Earth: The Social Construction of the Landsat Satellite System*

Donald MacKenzie, *An Engine, Not a Camera: How Financial Models Shape Markets*

Donald MacKenzie, *Inventing Accuracy: A Historical Sociology of Nuclear Missile Guidance*

Donald MacKenzie, *Knowing Machines: Essays on Technical Change*

Donald MacKenzie, *Mechanizing Proof: Computing, Risk, and Trust*

Maggie Mort, *Building the Trident Network: A Study of the Enrollment of People, Knowledge, and Machines*

Nelly Oudshoorn and Trevor Pinch, editors, *How Users Matter: The Co-Construction of Users and Technology*

Paul Rosen, *Framing Production: Technology, Culture, and Change in the British Bicycle Industry*

Susanne K. Schmidt and Raymund Werle, *Coordinating Technology: Studies in the International Standardization of Telecommunications*

Charis Thompson, *Making Parents: The Ontological Choreography of Reproductive Technology*

Dominique Vinck, editor, *Everyday Engineering: An Ethnography of Design and Innovation*